Lecture Notes in Mathematics

Edited by A. Dold, B. Eckmann and F. Takens

1433

Serge Lang   William Cherry

Topics in Nevanlinna Theory

Springer-Verlag

Berlin Heidelberg New York London
Paris Tokyo Hong Kong Barcelona

Authors

Serge Lang
William Cherry
Department of Mathematics, Yale University
Box 2155 Yale Station, New Haven, CT 06520, USA

Mathematics Subject Classification (1980): 30D35, 32A22, 32H30

ISBN 3-540-52785-0 Springer-Verlag Berlin Heidelberg New York
ISBN 0-387-52785-0 Springer-Verlag New York Berlin Heidelberg

© Springer-Verlag Berlin Heidelberg 1990
Printed in Germany

Printing and binding: Druckhaus Beltz, Hemsbach/Bergstr.
2146/3140-543210 – Printed on acid-free paper

PART ONE

LECTURES ON NEVANLINNA THEORY

by Serge Lang

CHAPTER I

NEVANLINNA THEORY IN ONE VARIABLE

CHAPTER II

EQUIDIMENSIONAL HIGHER DIMENSIONAL THEORY

PART TWO

NEVANLINNA THEORY OF COVERINGS

by William Cherry

CHAPTER III

NEVANLINNA THEORY FOR MEROMORPHIC FUNCTIONS ON COVERINGS OF C

CHAPTER IV

EQUIDIMENSIONAL NEVANLINNA THEORY ON COVERINGS OF C^n

PART ONE

LECTURES ON
NEVANLINNA THEORY

by Serge Lang

INTRODUCTION

These are notes of lectures on Nevanlinna theory, in the classical case of meromorphic functions (Chapter I) and the generalization by Carlson-Griffiths to equidimensional holomorphic maps $f\colon \mathbf{C}^n \to X$ where X is a compact complex manifold (Chapter II). Until recently, no special attention was paid to the significance of the error term in Nevanlinna's main inequality, see for instance Shabat's book [**Sh**]. In [**La 8**] I pointed to the existence of a structure to this error term and conjectured what could be essentially the best possible form of this error term in general. I also emphasized the importance of determining the best possible error term for each of the classical functions. I shall give a more detailed discussion of these problems in the introduction to Chapter I. In this way, new areas are opened in complex analysis and complex differential geometry. I shall also describe the way I was inspired by Vojta's analogy between Nevanlinna Theory and the theory of heights in number theory.

P.M. Wong used a method of Ahlfors to prove my conjecture in dimension 1 [**Wo**]. In higher dimension, there was still a discrepancy between his result and the one in [**La 8**], neither of which contained the other. By an analysis of Wong's proof, I was able to make a certain technical improvement at one point which leads to the desired result, conjecturally best possible in general. Using Wong's approach, I was also able to give the same type of structure to the error term in Nevanlinna's theorem on the logarithmic derivative. As a result, it seemed to me useful to give a leisurely exposition which might lead people with no background in Nevanlinna theory to some of the basic problems which now remain about the error term. The existence of these problems and the possibly rapid evolution of the subject in light of the new viewpoints made me wary of writing a book, but I hope these lecture notes will be helpful in the meantime, and will help speed up the development of the subject. They might very well be used as a continuation for a graduate course in complex analysis, also leading into complex differential geometry. Sections 1 and 2 of Chapter I provide a natural bridge, and Chapter I is especially well suited to be used in conjunction with a course in complex analysis, to give applications for the Poisson and Jensen formulas which are usually proved at the end of such a course.

I have not treated Cartan's theorem, giving a second main theorem for holomorphic maps $f: \mathbf{C} \to \mathbf{P}^n$, because I gave Cartan's proof in [**La 7**], in a self contained way, and it is still the shortest and clearest. However, at the end of Chapter II, I give one application of the techniques to one case of a map $f: \mathbf{C} \to Y$ into a possibly non compact complex manifold as an illustration of the techniques in this case. I also have not given the theory of derived curves, which introduced complications of multilinear algebra in Cartan and Ahlfors [**Ah**], and prevented seeing more clearly certain phenomena having to do with the error term, which form our main concern here. I also want to draw attention to Vojta's result for maps $f: \mathbf{C} \to \mathbf{P}^n$ into projective space. In Cartan, Ahlfors, and Schmidt's version (the number theoretic case), it is assumed that the image of f does not lie in any hyperplane. Vojta was able to weaken this assumption to the image of f not lying in a finite union of hyperplanes, which he determines explicitly as generalized diagonals [**Vo 2**]. This result gives substantial new insight into the structure of the "exceptional set" in the linear case. Ultimately, this and other advances will also have to be included into a more complete book account of the theory, as it is now developing.

Serge Lang

6

Acknowledgement

I want to thank Donna Belli for typing these lecture notes, and doing such a beautiful job of computer setting in a triumph of person over machine.

S.L.

TERMINOLOGY AND BASIC NOTATION

By **increasing** we shall mean weakly increasing throughout, so an increasing function is allowed to be constant. **Positive** will mean strictly positive.

The open **disc** of radius R centered at the origin is denoted $\mathbf{D}(r)$. The **circle** of radius r centered at the origin is $\mathbf{S}(r)$. The closed disc is $\overline{\mathbf{D}}(r)$.

In $\mathbf{C}^n$, the **ball** and **sphere** of radius r are denoted

$$\mathbf{B}(r), \ \overline{\mathbf{B}}(r), \ \text{and} \ \mathbf{S}(r)$$

respectively.

Let F_1, F_2 be two positive functions defined for all real numbers $\geq r_0$. We write

$$F_1 \ll F_2$$

to mean that $F_1 = O(F_2)$. We shall write

$$F_1 \gg\!\!\ll F_2$$

to mean that $F_1 \ll F_2$ and $F_2 \ll F_1$.

CHAPTER I

NEVANLINNA THEORY IN ONE VARIABLE

In the first part of this chapter we essentially follow Nevanlinna, as in his book [**Ne**]. The main difference lies in the fact that we are careful about the error term in Nevanlinna's main theorem. That this error term has an interesting structure was first brought up in [**La 8**], in analogy with a similar conjecture in number theory. Although Osgood [**Os**] did notice a similarity between the 2 in the Nevanlinna defect and the 2 in Roth's theorem, Vojta gave a much deeper analysis of the situation, and compared the theory of heights in number theory or algebraic geometry with the Second Main Theorem of Nevanlinna theory.

In [**La 2**] and [**La 3**] I defined a **type** for a number α to be an increasing function ψ such that

$$-\log\left|\alpha - \frac{p}{q}\right| - 2\log q \leq \log \psi(q)$$

for all but a finite number of fractions p/q in lowest form, $q > 0$. The **height** $h(p/q)$ is defined as

$$\log \max(|p|, |q|).$$

If p/q is close to α then $\log q$ has the same order of magnitude as the height, so $\log q$ is essentially the height in the above inequality. A theorem of Khintchine states that almost all numbers have type $\leq \psi$ if

$$\sum \frac{1}{q\psi(q)} < \infty.$$

The idea is that algebraic numbers behave like almost all numbers, although it is not clear a priori if Khintchine's principle will apply without any further restriction on the function ψ. Roth's theorem can be formulated as saying that an algebraic number has type $\leq q^\varepsilon$ for every $\varepsilon > 0$, and in the sixties I conjectured that this could be improved to having type $\leq (\log q)^{1+\varepsilon}$ in line with Khintchine's principle. Cf. [La 1], [La 3], [La 4] especially.[1] Thus for instance we would have the improvement of Roth's inequality

$$\left| \alpha - \frac{p}{q} \right| \geq \frac{C(\alpha, \varepsilon)}{q^2 (\log q)^{1+\varepsilon}}$$

which could also be written

$$-\log \left| \alpha - \frac{p}{q} \right| - 2 \log q \leq (1 + \varepsilon) \log \log q$$

for all but a finite number of fractions p/q. However, except for quadratic numbers that have bounded type, there is no known example of an algebraic number about which one knows that it is or is not of type $(\log q)^k$ for some number $k > 1$. It becomes a problem to determine the type for each algebraic number and for the classical numbers. For instance, it follows from Adams' work [Ad 1], [Ad 2] that e has type

$$\psi(q) = \frac{C \log q}{\log \log q}$$

with a suitable constant C, which is much better than the "probabilistic" type $(\log q)^{1+\varepsilon}$.

In light of Vojta's analysis, it occurred to me to transpose my conjecture about the "error term" in Roth's theorem to the context of Nevanlinna theory, in one and higher dimension. Transposing to the analytic context, it becomes a problem to determine the "type" of the

---

[1] Unknown to me until much later, similar conjectures were made by Bryuno [Br] and Richtmyer, Devaney and Metropolis [RDM], see [L-T 1] and [L-T 2].

classical meromorphic functions, i.e. the best possible error term in the second main theorem which describes the value distribution of the function. It is classical, and easy, for example, that e^z has bounded type, i.e. that the error term in the Second Main Theorem is $O(1)$. Two problems arise here:

- To determine for "almost all" functions (in a suitable sense) whether the type follows the pattern of Khintchine's convergence principle.

- To determine the specific type for each concrete classical function, using the specific special properties of each such function: $\wp, \theta, \Gamma, \zeta, J$, etc.

I am much indebted to Ye for an appendix exhibiting functions of type corresponding to a factor of $1 - \varepsilon$ in number theory. Until he gave these examples, I did not even know a function which did not have bounded type.

In [**La 8**], using the singular volume form of Carlson-Griffiths or a variation of it, I was not able to prove my conjecture exactly, with the correct factor of $1 + \varepsilon$ (I got only $3/2$ instead of 1).

P.M. Wong brought back to life a method which occurs in Ahlfors' original 1941 paper, and by this method he established my conjecture with the $1 + \varepsilon$. I pointed out to him that his method would also prove the desired result with an arbitrary type function satisfying only the convergence of the integral similar to the Khintchine principle. The role of the convergence principle becomes very clear in that method, which is given in the second part of this chapter. The method had also been tried improperly by Chern in the early sixties, and we shall have more to say on the technical aspects when we come to the actual theorems in §4. Ahlfors' method was obscured for a long time by other technical aspects of his paper, and I think Wong made a substantial contribution by showing how it could be applied successfully. I should also note that H. Wu also proved the conjecture with $1+\varepsilon$ (unpublished) by the "averaging method" of Ahlfors. But the method used by Wong

11

lent itself better to give the generalized version with the function ψ.

Developing fully the two problems mentioned above would create a whole new area of complex analysis, digging into properties of meromorphic functions in general, and of the classical functions in particular, which up to now have been disregarded.

I, §1. THE POISSON-JENSEN FORMULA AND THE NEVANLINNA FUNCTIONS

By a meromorphic function we mean a meromorphic function on the whole plane, so its zeros and poles form a discrete set. A meromorphic function on a closed set (e.g. the closed disc $\overline{\mathbf{D}}(R)$) is by definition meromorphic on some open neighborhood of the set.

Theorem 1.1. (Poisson formula) *Let f be holomorphic on the closed disc $\overline{\mathbf{D}}(R)$. Let z be inside the disc, and write $z = re^{i\varphi}$. Then*

$$f(z) = \int_0^{2\pi} f(Re^{i\theta}) \operatorname{Re} \frac{Re^{i\theta} + z}{Re^{i\theta} - z} \frac{d\theta}{2\pi}$$

$$= \int_0^{2\pi} f(Re^{i\theta}) \frac{R^2 - r^2}{R^2 - 2R\cos(\theta - \varphi) + r^2} \frac{d\theta}{2\pi}$$

$$= \int_0^{2\pi} \operatorname{Re} f(Re^{i\theta}) \frac{Re^{i\theta} + z}{Re^{i\theta} - z} \frac{d\theta}{2\pi} + iK \quad \text{for some real constant } K.$$

Proof: By Cauchy's theorem,

$$f(0) = \frac{1}{2\pi i} \int_{S_R} \frac{f(\zeta)}{\zeta} d\zeta = \int_0^{2\pi} f(Re^{i\theta}) \frac{d\theta}{2\pi}.$$

Let g be the automorphism of $\overline{\mathbf{D}}(R)$ which interchanges 0 and z. Then

$$f(z) = f \circ g(0).$$

12

We apply the above formula to $f \circ g$. We then change variables, and use the fact that $g \circ g = \mathrm{id}$, $\zeta = g(w)$, $d\zeta = g'(w)dw$. For $R = 1$,

$$g(w) = \frac{z - w}{1 - w\bar{z}}.$$

The desired formula drops out as in the first equation. The identity

$$\mathrm{Re}\frac{Re^{i\theta} + z}{Re^{i\theta} - z} = \frac{R^2 - r^2}{R^2 - 2R\cos(\theta - \varphi) + r^2}$$

is immediate by direct computation. The third equation comes from the fact that f and the integral on the right hand side of the third equation are both analytic in z, and have the same real part, so differ by a pure imaginary constant. This concludes the proof.

For $a \in \mathbf{D}(R)$ define

$$G_R(z, a) = G_{R,a}(z) = \frac{R^2 - \bar{a}z}{R(z - a)}.$$

Then $G_{R,a}$ has precisely one pole on $\overline{\mathbf{D}}(R)$ and no zeros. We have

$$|G_{R,a}(z)| = 1 \text{ for } |z| = R.$$

Theorem 1.2. (Poisson-Jensen formula) *Let f be meromorphic non constant on $\mathbf{D}(R)$. Then for any simply connected open subset of $\mathbf{D}(R)$ not containing the zeros or poles of f, there is a real constant K such that for z in the open set we have*

$$\log f(z) = \int_0^{2\pi} \log |f(Re^{i\theta})|\frac{Re^{i\theta} + z}{Re^{i\theta} - z}\frac{d\theta}{2\pi}$$
$$- \sum_{a \in \mathbf{D}(R)} (\mathrm{ord}_a f) \log G_R(z, a) + iK.$$

The constant K depends on a fixed determination of the logs.

13

Proof: Suppose first that f has no zeros or poles on $\mathbf{S}(R)$. Let

$$h(z) = f(z) \prod \left(\frac{R^2 - \bar{a}z}{R(z-a)} \right)^{\mathrm{ord}_a f}$$

where the product is taken for $a \in \mathbf{D}(R)$. Then h has no zero or pole on $\overline{\mathbf{D}}(R)$, and so $\log h$ is defined as a holomorphic function to which we can apply Theorem 1.1 to get the present formula. Then we use the fact that the log of a product is the sum of the logs plus a pure imaginary constant, on a simply connected open set, to conclude the proof in the present case.

Suppose next that f may have zeros and poles on $\mathbf{S}(R)$. Note that

$$\theta \longmapsto \log |f(Re^{i\theta})|$$

is absolutely integrable, because where there are singularities, they are like $\log |x|$ in a neighborhood of the origin $x = 0$ in elementary calculus. Let R_n be a sequence of radii having R as a limit. For R_n sufficiently close to R, the zeros and poles of f inside the disc of radius R_n are the same as the zeros and poles of f inside the disc of radius R, except for the zeros and poles lying on the circle $\mathbf{S}(R)$. The left hand side of the formula is independent of R_n. Let

$$\varphi_n(\theta) = \log |f(R_n e^{i\theta})| \qquad \text{and} \qquad \varphi(\theta) = \log |f(Re^{i\theta})|.$$

Then φ_n converges to φ. Outside small intervals around the zeros or poles of f on $\mathbf{S}(R)$, the convergence is uniform. Near the zeros and poles of f on the circle, the contribution of the integrals over small θ-intervals is small. Hence

$$\int_0^{2\pi} \log |f(R_n e^{i\theta})| \frac{Re^{i\theta} + z}{Re^{i\theta} - z} \frac{d\theta}{2\pi} \quad \text{converges to} \quad \int_0^{2\pi} \log |f(Re^{i\theta})| \frac{Re^{i\theta} + z}{Re^{i\theta} - z} \frac{d\theta}{2\pi},$$

thus proving the formula in general.

14

From the Poisson-Jensen formula, we deduce a slightly simpler relationship for the real parts, namely:

For all $z \in \mathbf{D}(R)$ which are not zeros or poles of f, we have

$$\log |f(z)| = \int_0^{2\pi} \log |f(Re^{i\theta})| \operatorname{Re} \frac{Re^{i\theta} + z}{Re^{i\theta} - z} \frac{d\theta}{2\pi}$$

$$- \sum_{a \in \mathbf{D}(R)} (\operatorname{ord}_a f) \log |G_R(z, a)|.$$

If 0 is not a zero or pole of f, then

$$\log |f(0)| = \int_0^{2\pi} \log |f(Re^{i\theta})| \frac{d\theta}{2\pi} - \sum_{\substack{a \in \mathbf{D}(r) \\ a \neq 0}} (\operatorname{ord}_a f) \log \left| \frac{R}{a} \right|.$$

In general, let $f(z) = c_f z^e + \cdots$ where $c_f \neq 0$ is the leading coefficient. Then

$$\log |c_f| = \int_0^{2\pi} \log |f(Re^{i\theta})| \frac{d\theta}{2\pi} - \sum_{a \neq 0} (\operatorname{ord}_a f) \log |R/a| - e \log R.$$

This last formula follows by applying the previous formula to the function $f(z)/z^e$.

Following Nevanlinna, we define the **counting functions**

$$n_f(0, R) = \text{number of zeros of } f \text{ in } \overline{\mathbf{D}}(R)$$
$$n_f(\infty, R) = \text{number of poles of } f \text{ in } \overline{\mathbf{D}}(R)$$
$$= n_f(R).$$

Thus $n_f(0, R) = n_{1/f}(\infty, R) = n_{1/f}(R)$. We also define:

$$N_f(0, R) = \sum_{\substack{a \in \mathbf{D}(R) \\ a \neq 0, f(a) = 0}} (\operatorname{ord}_a f) \log \left| \frac{R}{a} \right| + (\operatorname{ord}_0 f) \log R$$

$$N_f(\infty, R) = N_{1/f}(0, R).$$

15

We may rewrite Jensen's formula with the above notation:

Theorem 1.3 *Let f be meromorphic on $\overline{D}(R)$. Then*

$$\int\limits_0^{2\pi} \log|f(Re^{i\theta})|\frac{d\theta}{2\pi} = \log|c_f| + N_f(0,R) - N_f(\infty,R).$$

If $a \in \mathbf{C}$ we define

$$n_f(a,r) = n_{f-a}(0,r) \text{ and } N_f(a,r) = N_{f-a}(r).$$

Proposition 1.4 *Let $n_f(r)$ and $N_f(r)$ denote $n_f(0,r)$ and $N_f(0,r)$. Then*

$$N_f(R) = \int\limits_0^R [n_f(t) - n_f(0)]\frac{dt}{t} + n_f(0)\log R.$$

Proof: The function

$$t \mapsto n_f(t) - n_f(0)$$

is a step function which is equal to 0 for t sufficiently close to 0. If we decompose the interval $[0, R]$ into subintervals whose end points are the jumps in the absolute values of the zeros of f, integrate over each such interval where the integrand is constant, and take the sum, then the formula of the proposition drops out.

Remark: If $R > 1$ we have $N_f(a, R) \geq 0$. If f has no zero at the origin, then $N_f(0, R) \geq 0$ for all $R > 0$. The only possibly non positive contribution to N_f is the term with $n_f(0, R)\log R$ with $0 < R < 1$. For simplicity, we shall often assume that f has no zero or pole at the origin, to get rid of this term.

For $\alpha > 0$ we define

$$\log^+ \alpha = \max(0, \log \alpha).$$

Proposition 1.5 *Let $b \in \mathbf{C}$. Then*

$$\int_0^{2\pi} \log |b - e^{i\theta}| \frac{d\theta}{2\pi} = \log^+ |b|.$$

This is an immediate consequence of Theorem 1.3, but can also be proved directly using the mean value property of harmonic functions. Indeed, if $|b| > 1$ then $\log |b - z|$ for $|z| < 1 + \varepsilon$ is harmonic, and $\log^+ |b| = \log |b|$, so the formula is true by the mean value property for harmonic functions. If $|b| < 1$, then subtracting $\log |b|$ from the left hand side, replacing θ by $-\theta$, and using the first part of the proof shows that the integral comes out to be 0, which is $\log^+ |b|$ again. Finally for $|b| = 1$, the formula follows by continuity of each side in b, and the absolute convergence of the integral on the left. This proves the proposition.

Remark: Proposition 1.5 shows that the expression $\log^+ |b|$ is in some sense natural.

As usual, we let the projective line $\mathbf{P}^1$ consist of $\mathbf{C}$ and a point ∞. Points of $\mathbf{P}^1$ can be identified with equivalence classes of pairs (w_1, w_2) with complex numbers w_1, w_2 not both 0, and

$$(w_1, w_2) \sim (cw_1, cw_2) \quad \text{if} \quad c \neq 0.$$

Each point of $\mathbf{P}^1$ has an affine representative $(w, 1)$ or $\infty = (1, 0)$ with $w \in \mathbf{C}$. A meromorphic function f can be identified with a holomorphic map into $\mathbf{P}^1$. Namely we write $f = f_1/f_0$ where f_1, f_0 are entire functions without common zeros, which can always be done by the Weierstrass factorization theorem. Then we obtain a holomorphic map into $\mathbf{P}^1$ given in terms of coordinates by

$$z \mapsto (f_0(z), f_1(z)).$$

We define a "distance" on $\mathbf{P}^1$ between two points w, w' to be

$$\|w, w'\|^2 = \frac{|w - w'|^2}{(1 + |w|^2)(1 + |w'|^2)}.$$

The formula makes sense if $w, w' \in \mathbf{C}$. If one but not both of $w, w' = \infty$, then we let

$$\|w, \infty\|^2 = \|\frac{1}{w}, 0\|^2 = \frac{1}{(1 + |w|^2)}.$$

We let $\|0, \infty\| = 1$. This distance satisfies the triangle inequality, as one can show by identifying it with what is sometimes called the "chordal distance" under stereographic projection, but we don't go into this here. Observe that if $w \neq w'$ then

$$0 < \|w, w'\| \leq 1, \quad \text{and} \quad < 1 \text{ except for antipodal points.}$$

Furthermore, the function $(w, w') \mapsto \|w, w'\|^2$ is a C^∞ (even real analytic) function on $\mathbf{P}^1 \times \mathbf{P}^1$.

Let f be a non-constant meromorphic function and let $a \in \mathbf{P}^1$. We define the **mean proximity** function to be

$$m_f(a, r) = \int\limits_0^{2\pi} -\log \|f(re^{i\theta}), a\| \frac{d\theta}{2\pi}.$$

The function $z \mapsto -\log \|f(z), a\|$ is a convenient normalization for the more general notion of a **Weil function**. Given $a \in \mathbf{P}^1$, we define a **Weil function associated with** a to be a continuous map

$$\lambda_a : \mathbf{P}^1 - \{a\} \longrightarrow \mathbf{R}$$

having the property that in some open neighborhood of a there exists a continuous function α such that

$$\lambda_a(z) = -\log |z - a| + \alpha(z).$$

The difference between two Weil functions is a continuous function on $\mathbf{P}^1$, which is bounded. Another way of normalizing a Weil function is to take

$$\lambda_a(z) = \log^+ \frac{1}{|z - a|} \qquad \text{if} \quad a \neq \infty$$

$$\lambda_a(z) = \log^+ |z| \qquad \text{if} \quad a = \infty.$$

However, the previous normalization is more natural and more symmetric in many respects. Given any Weil function, one can define a corresponding proximity function by

$$m_f(\lambda_a, r) = \int_0^{2\pi} \lambda_a(f(re^{i\theta})) \frac{d\theta}{2\pi}.$$

It differs from $m_f(a, r)$ by a bounded function. Our normalization of m_f insures that

$$m_f(a, r) \geq 0.$$

For simplicity we shall assume throughout that

$$f(0) \neq 0, \infty \text{ and } f'(0) \neq 0.$$

We define the **height** to be the function $T_f \colon \mathbf{R}_{>0} \to \mathbf{R}$ given by

$$T_f(\infty, r) = T_f(r) = m_f(\infty, r) + N_f(\infty, r) + \log \|f(0), \infty\|,$$

and more generally if $f(0) \neq a$

$$T_{f,a}(r) = m_f(a, r) + N_f(a, r) + \log \|f(0), a\|.$$

Theorem 1.6 (First Main Theorem). *The function $T_f(a, r)$ is independent of $a \in \mathbf{P}^1$, provided $f(0) \neq a$.*

Proof: Applying Jensen's formula to $f - a$, and $a \neq \infty$, we get

$$-\int_0^{2\pi} \log |f(re^{i\theta}) - a| \frac{d\theta}{2\pi} = \log |f(0) - a| - N_{f-a}(0, r) + N_{f-a}(\infty, r).$$

19

But the poles of $f - a$ are the same as the poles of f, so $N_{f-a}(\infty, r) = N_f(\infty, r)$. On the other hand, by definition of the symbols,

$$T_{f,a}(r) = -\int_0^{2\pi} \log \|f(re^{i\theta}), a\| \frac{d\theta}{2\pi} + N_{f,a}(r) + \log \|f(0), a\|$$

$$= -\int_0^{2\pi} \log |f(re^{i\theta}) - a| \frac{d\theta}{2\pi} + \frac{1}{2} \int_0^{2\pi} \log(1 + |f(re^{i\theta})|^2) \frac{d\theta}{2\pi}$$

$$+ \frac{1}{2} \log(1 + |a|^2) + N_{f-a}(0, r) + \log \|f(0), a\|.$$

Using the definitions and the above Jensen formula, it is then immediate to verify that the right hand side is equal to $T_f(\infty, r)$, thus concluding the proof.

In light of the theorem, one simply writes $T_f(r)$ for the Nevanlinna height. Observe that the only non positive contribution to $T_f(r)$ is the constant term involving $f(0)$. Note that as a function of $a \in \mathbf{P}^1$, this term is absolutely integrable on $\mathbf{P}^1$ with respect to any C^∞ form.

I, §2. THE DIFFERENTIAL GEOMETRIC DEFINITIONS AND GREEN-JENSEN'S FORMULA

Let $f: \mathbf{C} \to \mathbf{P}^1$ be a holomorphic map, which we identify with a meromorphic function. We define the **Fubini-Study** form ω on $\mathbf{P}^1$ to be the form given in terms of an affine coordinate w by

$$\omega = \frac{\sqrt{-1}}{2\pi} \frac{dw \wedge d\bar{w}}{(1 + |w|^2)^2}.$$

We let the **euclidean form** Φ on $\mathbf{C}$ be the form

$$\Phi = \frac{\sqrt{-1}}{2\pi} dz \wedge d\bar{z} = 2r dr \frac{d\theta}{2\pi}.$$

20

Then
$$\text{letting} \quad \gamma_f = \frac{|f'|^2}{(1+|f|^2)^2}, \quad \text{we have} \quad f^*\omega = \gamma_f \Phi.$$

Note that γ_f is C^∞ (actually real analytic). By the Weierstrass factorization theorem, we can write $f = f_1/f_0$ where f_1, f_0 are entire functions without common zero. We let

$$W = W(f_0, f_1) = f_0 f_1' - f_0' f_1$$

be the **Wronskian**. Then we can write

$$\gamma_f = \frac{|W|^2}{\left(|f_0|^2 + |f_1|^2\right)^2}.$$

Note that $|f_0|^2 + |f_1|^2$ is a positive C^∞ function.

Let ∂ and $\bar{\partial}$ be the usual differential operators, so for instance for a C^∞ function α we have

$$\partial \alpha = \frac{\partial \alpha}{\partial z} dz \quad \text{and} \quad \bar{\partial} \alpha = \frac{\partial \alpha}{\partial \bar{z}} d\bar{z}.$$

We let as usual

$$d = \partial + \bar{\partial} \quad \text{and} \quad d^c = \frac{1}{2\pi} \frac{\partial - \bar{\partial}}{2i}.$$

Then

$$dd^c = \frac{\sqrt{-1}}{2\pi} \partial \bar{\partial}.$$

Let g be a meromorphic function. Then outside the zeros and poles of g, we have

$$\partial \bar{\partial} g = 0$$

because of the Cauchy-Riemann equation $\bar{\partial} g = 0$. Also

$$dd^c \log |g|^2 = 0$$

21

outside the zeros and poles of g, because $\log(g\bar{g}) = \log|g|^2$ is defined and equal to $\log g + \log \bar{g}$ in any simply connected open set, up to an additive constant, and again applying $\partial\bar{\partial}$ kills this expression by Cauchy-Riemann. We emphasize that when we write a differential operator applied to a function with singularities (zeros and poles) as above, we mean the operator applied to the function on the complement of the singularities. An appropriate notation will be developed later when we need to consider the singularities explicitly.

We note that in polar coordinates (r, θ) the operator d^c has the form

$$d^c\alpha = \frac{1}{2}r\frac{\partial\alpha}{\partial r}\frac{d\theta}{2\pi} - \frac{1}{4\pi}\frac{1}{r}\frac{\partial\alpha}{\partial\theta}dr.$$

The term with dr disappears when we restrict the function to a circle centered at the origin.

Proposition 2.1 *We have (outside the singularities)*

$$\gamma_f\Phi = dd^c\log(1 + |f|^2) = dd^c\log(|f_0|^2 + |f_1|^2)$$
$$= -\frac{1}{2}dd^c\log\gamma_f$$

Proof: Using the fact that d, d^c are derivations, one has

$$dd^c\log(1 + u) = \frac{dd^cu}{1 + u} - \frac{du \wedge d^cu}{(1 + u)^2}.$$

But, $u^2dd^c\log u = udd^cu - du \wedge d^cu$. Applying this to $u = |f|^2$ and $dd^c\log|f|^2 = 0$ will give the formula. We use the fact that

$$dd^c\log|f_0|^2 = 0$$

for the second equality sign.

We shall apply Stokes' theorem in the following context.

Proposition 2.2. *Let α be a C^2 function except for a discrete set of singularities. Suppose that α has no singularity on the circle S(t).*

Let Z be the set of singularities in $\mathbf{D}(t)$, and let $S(Z, \varepsilon)$ be the (formal) sum of small circles around these singularities. Then

(1)
$$\int_{\mathbf{D}(t)} dd^c \alpha = \int_{\mathbf{S}(t)} d^c \alpha - \lim_{\varepsilon \to 0} \int_{S(Z, \varepsilon)} d^c \alpha.$$

If α has no singularities, i.e. Z is empty, then

(2)
$$\int_{\mathbf{D}(t)} dd^c \alpha = \int_{\mathbf{S}(t)} d^c \alpha.$$

On the other hand, if $\alpha = \log |g|^2$ where g is holomorphic, then

(3)
$$\lim_{\varepsilon \to 0} \int_{S(Z, \varepsilon)} d^c \log |g|^2 = n_g(0, t).$$

Proof: The first formula is simply the Stokes-Green formula in the plane. For the third formula, we are reduced to proving it for a circle around a single zero or pole, so without loss of generality, let us assume that the point is the origin. We then have to prove

$$\lim_{\varepsilon \to 0} \int_{\mathbf{S}(\varepsilon)} d^c \log |g|^2 = \mathrm{ord}_0(g).$$

Let $k = \mathrm{ord}_0(g)$. We can write $g\bar{g} = r^{2k} h(r, \theta)$ where h is C^∞ and > 0. The operator $d^c \log$ transforms multiplication to addition, so we are reduced to replacing $|g|^2$ by r^{2k}. In this case, using the expression for d^c in polar coordinates concludes the proof.

In the next result, we shall need to differentiate under an integral sign, and we also need continuity with respect to parameters. There is a basic set of conditions under which these properties hold, carried out in detail for instance in my *Real Analysis*. We recall these here in our context.

Let α be a C^2 function on $\mathbf{C}$ except for a discrete set of points, and assume α continuous at 0. We consider the following conditions on α, which will insure continuity and differentiability (**CD**):

CD 1. *For each r, the function $\theta \mapsto \alpha(re^{i\theta})$ is absolutely integrable.*

CD 2. *Given $r_0 > 0$ there exists an open interval around r_0 and an L^1 function α_1 of θ alone such that*

$$|\alpha(re^{i\theta})| \leq \alpha_1(\theta) \quad \text{for all} \quad r \quad \text{in the interval.}$$

CD 3. *Given $r_0 > 0$ there exists an open interval around r_0 and an L^1 function α_2 of θ alone such that*

$$|\frac{\partial}{\partial r}\alpha(re^{i\theta})| \leq \alpha_2(\theta) \text{ for all } r \text{ in the interval.}$$

Proposition 2.3. Green-Jensen Formula *Let α be a C^2 function except at a discrete set of points, and continuous at 0. Assume that α also satisfies the three **CD** conditions. Then*

$$r \mapsto \int\limits_0^{2\pi} \alpha(re^{i\theta})\frac{d\theta}{2\pi}$$

is continuous for $r \geq 0$, and

$$\int\limits_0^r \frac{dt}{t} \int\limits_{\mathbf{D}(t)} dd^c\alpha + \int\limits_0^r \frac{dt}{t} \lim_{\varepsilon \to 0} \int\limits_{S(Z,\varepsilon)(t)} d^c\alpha = \frac{1}{2}\int\limits_0^{2\pi} \alpha(re^{i\theta})\frac{d\theta}{2\pi} - \frac{1}{2}\alpha(0).$$

Proof: The proof is immediate by integrating the previous relation, and taking into account the polar expression for d^c. We can take $\partial/\partial t$ outside the integral sign, namely

$$\int\limits_0^r \frac{dt}{t} \int\limits_{S(t)} d^c\alpha = \int\limits_0^r \frac{dt}{t} \int\limits_0^{2\pi} \frac{1}{2}t\frac{\partial}{\partial t}\alpha(te^{i\theta})\frac{d\theta}{2\pi} - \frac{1}{2}\int\limits_0^r \frac{\partial}{\partial t}(\int\limits_0^{2\pi} \alpha(te^{i\theta})\frac{d\theta}{2\pi})dt$$

so the proposition falls out by Stokes' formula and the standard criteria for differentiating under the integral sign.

Remark: We are avoiding here the language of distribution, which actually would not shorten what we are doing. In that language, if we denote by $[\alpha]$ the distribution associated with the function α, then the **Green-Jensen formula** may be stated in the form

$$\int_0^r \frac{dt}{t} \int_{\mathbf{D}(t)} dd^c[\alpha] = \frac{1}{2} \int_0^{2\pi} \alpha(re^{i\theta}) \frac{d\theta}{2\pi} - \frac{1}{2}\alpha(0).$$

For our purposes, we simple **define** $dd^c[\alpha]$ to be the functional such that

$$\int \frac{dt}{t} \int_{\mathbf{D}(t)} dd^c[\alpha] = \int_0^r \frac{dt}{t} \int_{\mathbf{D}(t)} dd^c\alpha + \mathrm{Sing}_\alpha(r),$$

where

$$\mathrm{Sing}_\alpha(r) = \int_0^r \frac{dt}{t} \lim_{\varepsilon \to 0} \int_{\mathbf{S}(Z,\varepsilon)(t)} d^c\alpha.$$

We call the first integral with $dd^c\alpha$ the **regular part** of $dd^c[\alpha]$, and we call $\mathrm{Sing}_\alpha(r)$ the **singular part**. In practice, the conditions under which we apply the formula are varied, and require a determination of the singular part, which sometimes will contribute to the formula in an essential way, and sometimes will be 0. Each time one must determine just what it is, so using "distributions" in a abstract nonsense sort of way does not help. The singular part may be 0 for different reasons. First, if the function α is C^∞, or even C^2, then the singular part is 0 because the limit as $\varepsilon \to 0$ is 0. But the singularities of α may be weak enough so that the limit is still 0, and examples of this phenomenon will be given below. On the other hand, when the singularities are like poles, then they contribute to the expression in an essential way, as in Proposition 2.2.

25

We shall give important examples of the Green-Jensen formula. First comes an alternative definition of the height.

Theorem 2.4 (Ahlfors-Shimizu). *Let f be meromorphic. Then*

$$T_f(r) = \int_0^r \frac{dt}{t} \int_{D(t)} \gamma_f \Phi.$$

Proof: In Proposition 2.3, let

$$\alpha = \log(1 + |f|^2).$$

Let $f = f_1/f_0$ where f_1, f_0 are holomorphic without common zeros. Let z_j be the zeros of f_0, that is the poles of f in the disc $D(t)$. By Proposition 2.2,

$$\lim_{\varepsilon \to 0} \int_{S(z,\varepsilon)} d^c \alpha = \operatorname{ord}_{z_j}(f_0),$$

because

$$d^c \log(1 + |f|^2) = d^c \log(|f_0|^2 + |f_1|^2) - d^c \log|f_0|^2,$$

and the smooth part contributes 0 to the limit. Thus the singular part contributes the number of poles of f in the disc, and its integral against dt/t contributes $N_f(\infty, r)$. On the other hand, the other part coming from the circle of radius r is precisely

$$m_f(\infty, r) + \log \|f(0), \infty\|,$$

thus proving the theorem.

Integrals of the form

$$\int_0^r \frac{dt}{t} \int_{D(t)} \alpha \Phi$$

will be encountered in a fundamental way, and will be given a name in the next section.

As an application of some of the above computations, we give formulations of special cases in terms of the Nevanlinna functions.

Corollary 2.5 *For f meromorphic non-constant, we have:*

$$T_f(r) = \int_0^r \frac{dt}{t} \int_{\mathbf{D}(t)} dd^c \log(1 + |f|^2)$$

$$N(f, a, r) = \int_0^r \frac{dt}{t} \int_{\mathbf{D}(t)} dd^c[\log|f - a|^2]$$

$$T_f(r) - N(f, a, r) = -\int_0^r \frac{dt}{t} \int_{\mathbf{D}(t)} dd^c[\log \|f, a\|^2].$$

Proof: The first formula for $T_f(r)$ comes from Proposition 2.1 and Theorem 2.4. The formula for $N(f, a, r)$ is a special case of Proposition 2.2(3). The last formula for the difference comes from the definition of $\|f, a\|^2$ and the fact that the operator $dd^c \log$ transforms multiplication into addition.

Next we give an example such that the function has singularities, but these singularities are sufficiently mild so that the singular term in the Green-Jensen formula vanishes.

Proposition 2.6 *Let $0 < \lambda$ and let $h = \|f, a\|^{2\lambda}$. Then*

$$\int_0^r \frac{dt}{t} \int_{\mathbf{D}(t)} dd^c \log(1 + h)$$

$$= \frac{1}{2} \int_0^{2\pi} \log(1 + h(re^{i\theta})) \frac{d\theta}{2\pi} - \frac{1}{2} \log(1 + h(0)).$$

Proof: The function h is C^∞ except possibly at the zeros of $f - a$ or the poles of f if $a = \infty$. If no zero or pole lies on $\mathbf{S}(t)$, then we shall prove that

$$\int_{\mathbf{D}(t)} dd^c \log(1+h) = \int_{\mathbf{S}(t)} d^c \log(1+h).$$

Integrating this against dt/t from 0 to r just as in the smooth case of Theorem 2.2 then proves the proposition.

For simplicity, we assume that f is holomorphic, and we have to show that the term coming from the singularities in Stokes' theorem give zero contribution.

Let $\beta = \log(1+h)$, so β is C^∞ except at the set of zeros of $f - a$, which we denote by Z. We let

$$S_\varepsilon(Z)(r) = \text{circles of radius } \varepsilon \text{ around each point of } Z$$
$$\text{lying inside } \mathbf{D}(r).$$

If $f - a$ has no zero on the circle $\mathbf{S}(t)$, then Stokes' theorem shows that

$$\int_{\mathbf{S}(t)} d^c\beta = \int_{\mathbf{D}(t)} dd^c\beta + \lim_{\varepsilon \to 0} \int_{S_\varepsilon(Z)(t)} d^c\beta.$$

For our special β, we claim that the limit on the right is 0. We are reduced to proving this limit in the neighborhood of each zero of $f - a$. Then in such a neighborhood, let z be a complex coordinate. Locally we are reduced to proving that if g is holomorphic at the origin, and

$$\beta = \log(1 + |g|^{2\lambda}\alpha_1)$$

where α_1 is > 0 and C^∞, then

$$\lim_{\varepsilon \to 0} \int_{r=\varepsilon} r \frac{\partial}{\partial r} \log(1 + |g|^{2\lambda}\alpha_1) \frac{d\theta}{2\pi} = 0.$$

28

We can rewrite $|g|^{2\lambda}a_1 = r^{2m\lambda}\alpha(r,\theta)$ where α is > 0 and C^∞, $m > 0$. Then

$$r\frac{\partial}{\partial r}\log(1 + r^{2m\lambda}\alpha(r,\theta)) = \frac{r2m\lambda r^{2m\lambda-1}\alpha(r,\theta) + r^{2m\lambda+1}\partial\alpha/\partial r}{1 + r^{2m\lambda}\alpha(r,\theta)}.$$

Setting $r = \varepsilon$ and letting $\varepsilon \to 0$ immediately shows that the desired limit is 0, thus proving the proposition.

I, §3. SOME CALCULUS LEMMAS

The expression of Theorem 2.4 will now be viewed as a transform, and some of its properties will depend only on calculus.

Lemma 3.1. *Let F be a positive increasing function defined for $r > 0$ with piecewise continuous derivative. Suppose there exists r_1 such that $F(r_1) \geq e$, say. Let ψ be a positive increasing function such that*

$$\int_e^\infty \frac{1}{u\psi(u)}du = b_0(\psi)$$

is finite. Then we have the inequality

$$F'(r) \leq F(r)\psi(F(r))$$

for $r \geq r_1$ outside a set of measure $\leq b_0(\psi)$.

Proof: The measure of the set of numbers $r \geq r_1$ such that

$$F'(r) \geq F(r)\psi(F(r))$$

is bounded by

$$\int_{r_1}^\infty \frac{F'(r)}{F(r)\psi(F(r))}dr \leq \int_e^\infty \frac{1}{u\psi(u)}du = b_0(\psi),$$

which proves the lemma.

Note: The calculus lemma was stated and proved in that generality by Nevanlinna. However, Nevanlinna and subsequent contributors to the field instead of keeping the function arbitrary, subject only to the convergent integral, specialized this function, and thereby prevented seeing clearly its formal role in the theory, analogous to the Khintchine convergence principle in diophantine approximations. I suggested keeping an arbitrary ψ systematically and defined the error term function a priori as follows:

Let F be a positive increasing function of class C^1 such that $r \mapsto rF'(r)$ is positive increasing. Let ψ be positive increasing, and let r, c be positive numbers. We define the **error term function**

$$S(F, c, \psi, r) = \log\{F(r)\psi(F(r))\psi(cr\,F(r)\psi(F(r)))\}$$
$$= \log F(r) + \log \psi(F(r)) + \log \psi(cr\,F(r)\psi(F(r))).$$

Note that S is monotone in all its four variables.

In practice, we may take ψ to be a slowly growing function, so that for instance

$$\psi(u) = (\log u)^{1+\epsilon} \text{ with } \epsilon > 0.$$

Then the measure of the exceptional set in Lemma 3.1 is bounded by $1/\epsilon$. Furthermore, with such ψ we see that the error term is of type

$$S = \log F(r) + \log \log \text{ terms.}$$

Lemma 3.2. *Let F be a function of class C^2 defined on $(0, \infty)$. Assume that both $F(r)$ and $rF'(r)$ are positive increasing functions of r, and that there exists $r_1 \geq 1$ such that*

$$F(r_1) \geq e.$$

Let $b_1(F)$ be the smallest number $b_1 \geq 1$ such that

$$b_1 r F'(r) \geq e \quad for \quad r \geq 1.$$

(Such a number trivially exists.) Then for all $r \geq r_1$ outside a set of measure $\leq 2b_0(\psi)$, and all $b_1 \geq b_1(F)$ we have

$$\frac{1}{r}\frac{d}{dr}\left(r\frac{dF}{dr}\right) \leq F(r)\psi(F(r))\psi(b_1 r F(r)\psi(F(r))),$$

and so

$$\log \frac{1}{r}\frac{d}{dr}\left(r\frac{dF}{dr}\right) \leq S(F, b_1, \psi, r).$$

Proof: We apply Lemma 3.1 twice, first to $b_1 r F'(r)$ and then to $F(r)$ to get the desired inequality with $b_1 = b_1(F)$. We can then use any other $b_1 \geq b_1(F)$ by the monotonicity of S.

If we choose $b_1 = b_1(F)$ in Lemma 3.2, then we sometimes omit b_1 from the notation, and write simply $S(F, \psi, r)$. But in some cases, we must view b_1 as a separate variable which we bound in its own right uniformly for a family of functions F.

Similarly, we let $r_1(F)$ be the smallest number $r_1 \geq 1$ such that

$$F(r_1) \geq e.$$

Note that if $F \leq G$ and $F' \leq G'$ then $r_1(F) \geq r_1(G)$ and $b_1(F) \geq b_1(G)$.

In the sequel we assume throughout that ψ satisfies the conditions as in Lemma 3.1, so ψ is positive increasing, and with the convergent integral.

Smoothness properties of ψ will be irrelevant. Just to fix ideas, one might assume ψ continuous or piecewise continuous, just to make the integral have a naive sense.

We shall now describe how to construct a function F to which we apply Lemmas 3.1 and 3.2.

We let α be a function on $\mathbf{C}$ such that:

(a) α is continuous and > 0 except at a discrete set of points;

(b) for each $r > 0$, the integral

$$\int_0^{2\pi} \alpha(r,\theta) \frac{d\theta}{2\pi}$$

is absolutely convergent, and gives a continuous function of r.

We recall the **euclidean form**

$$\Phi = \frac{\sqrt{-1}}{2\pi} dz \wedge d\bar{z} = 2r \, dr \, d\theta / 2\pi,$$

and define the **height transform** F_α of α by

$$F_\alpha(r) = \int_0^r \frac{dt}{t} \int_{D(t)} \alpha \Phi \qquad \text{for } r > 0.$$

Note that F_α is differentiable and has positive derivative, so is strictly increasing. We shall need to assume here a condition which is satisfied in practice, namely:

(c) There is some number $r_1 \geq 1$ such that $F_\alpha(r_1) \geq e$.

Lemma 3.3 *The function F_α is of class C^2 and we have*

$$\frac{1}{r}\frac{\partial}{\partial r}\left(r\frac{\partial}{\partial r}F_\alpha\right) = 2\int_0^{2\pi} \alpha(r,\theta)\frac{d\theta}{2\pi}.$$

Proof: First,

$$r\frac{\partial F}{\partial r} = 2\int_0^r \int_0^{2\pi} \alpha(t,\theta) t \, dt \frac{d\theta}{2\pi}.$$

32

Differentiating once more yields the desired formula.

We note that $rF'_\alpha(r)$ is an increasing function of r, and is > 0 for $r > 0$ in light of the assumption (a) on α. In particular, there exists $b_1 > 0$ such that

$$b_1 r F'_\alpha(r) \geq e \text{ for } r \geq 1,$$

and so the assumptions of Lemma 3.2 are satisfied.

The function $r \mapsto S(F, \psi, r)$ will be viewed as an error term, and depends on F and ψ. If $F = F_\alpha$ then this function S depends on α and ψ. We shall fix ψ throughout the applications, but we shall apply the inequality to various choices of α.

Putting Lemma 3.2 and Lemma 3.3 together we get:

Lemma 3.4. *Given α satisfying (a), (b), (c),we have*

$$\log \int_0^{2\pi} \alpha(r, \theta) \frac{d\theta}{2\pi} \leq S(F_\alpha, \psi, r)$$

for all $r \geq r_1(F_\alpha)$ outside a set of measure $\leq 2b_0(\psi)$.

Lemma 3.4 will be applied by taking a log in and out of an integral. We recall that result here for the convenience of the reader.

Lemma 3.5 *Let X be a measured space with positive measure μ and total measure 1. Let α be a real valued function ≥ 0 such that $\log \alpha$ is integrable. Then*

$$\int_X \log \alpha \, d\mu \leq \log \int_X \alpha d\mu.$$

Proof: Let

$$\int_X \alpha \, d\mu = c.$$

33

If $c = 0$, then α is 0 almost everywhere, $\log \alpha = -\infty$ almost everywhere, and the result is trivial. Suppose $c > 0$. We have to show

$$\int_X \log \alpha \, d\mu \leq \int_X (\log c) \, d\mu,$$

or equivalently,

$$\int_X \log(\alpha/c) \, d\mu \leq 0.$$

Write $\alpha/c = 1 + x$ and use $\log(1 + x) \leq x$ for $x > -1$. The desired inequality drops out from the definitions.

I, §4. RAMIFICATION AND SECOND MAIN THEOREM

We shall now apply §3 to the special case of a meromorphic function f, viewed as a holomorphic map of $\mathbf{C}$ into $\mathbf{P}^1$. Recall that

$$f^*\omega = \gamma_f \Phi.$$

Then from the definitions, the height transform of γ_f is precisely T_f, that is

$$F_{\gamma_f}(r) = T_f(\infty, r) = T_f(r).$$

We now define the ramification. Suppose $f(z_0) = a$. With $e \geq 1$, write

$$f(z) = c_0 + c_e(z - z_0)^e + \text{ higher terms with } c_e \neq 0 \text{ if } a \neq \infty$$
$$f(z) = c_e(z - z_0)^{-e} + \text{ higher terms with } c_e \neq 0 \text{ if } a = \infty.$$

We call e the **ramification index**, and we let $e - 1$ be the **order of the ramification divisor** of f at z_0. If $f = f_1/f_0$ where f_0, f_1 are entire without common zero, then it is immediately verified that

$$e - 1 = \text{order of } f_0 f_1' - f_0' f_1 \text{ at } z_0 = \text{order of } W(f_0, f_1).$$

34

Thus the zeros of the Wronskian, with multiplicities, define the ramification divisor of f. We define

$$n_{f,\mathrm{Ram}}(r) = n_W(0,r) \quad \text{and} \quad N_{f,\mathrm{Ram}}(r) = N_W(0,r).$$

Theorem 4.1. *Let f be a meromorphic function, non-constant, such that $f(0) \neq 0, \infty$ and $f'(0) \neq 0$. Then*

$$N_{f,\mathrm{Ram}}(r) - 2T_f(r) \leq \frac{1}{2}S(T_f, \psi, r) - \frac{1}{2}\log \gamma_f(0)$$

for $r \geq r_1(T_f)$ outside a set of measure $\leq 2b_0(\psi)$.

Proof: In Proposition 2.3 we take

$$\alpha = \log \gamma_f$$

where γ_f is the C^∞ function given by

$$\gamma_f = \frac{|f'|^2}{(1+|f|^2)^2} = \frac{|W|^2}{(|f_0|^2 + |f_1|^2)^2}.$$

Then

$$dd^c[\log \gamma_f] = dd^c[\log |W|^2] - 2dd^c \log(|f_0|^2 + |f_1|^2).$$

By Proposition 2.1 and the Ahlfors-Shimizu expression for the height in Proposition 2.4, and by Proposition 2.3 which gives us the singular part for the term with W, we find

$$N_{f,\mathrm{Ram}}(r) - 2T_f(r) = \int_0^r \frac{dt}{t} \int_{\mathbf{D}(t)} dd^c[\log \gamma_f]$$

$$= \frac{1}{2}\int_0^{2\pi} \log \gamma_f(re^{i\theta})\frac{d\theta}{2\pi} - \frac{1}{2}\log \gamma_f(0)$$

$$[\text{by Lemma 3.5}] \leq \frac{1}{2}\log \int_0^{2\pi} \gamma_f(re^{i\theta})\frac{d\theta}{2\pi} - \frac{1}{2}\log \gamma_f(0).$$

35

Since $F_{\gamma_f} = T_f$ we can apply Lemma 3.4 to conclude the proof.

Note: In the future, it will be convenient to give a name to the constant appearing on the right hand side, so we let

$$b_2(f) = -\frac{1}{2} \log \gamma_f(0).$$

Note: The argument for the proof of Theorem 4.1 goes back to Nevanlinna, who used the error term $O(\log rT_f(r))$, as did subsequent authors. The emphasis on trying to find a "best possible" error term stems from [**La 8**], and getting the factor $1/2$ in front of the error term stems from [**La 8**]. The error term as given here is due to Lang.

We shall extend Theorem 4.1 to the case when f can approximate a finite set of points, by measuring this closeness.

Theorem 4.2. *Let $q \geq 1$. Let $a_1, \ldots, a_q$ be a finite set of distinct points of $\mathbf{P}^1$, and let $f \colon \mathbf{C} \to \mathbf{P}^1$ be a non-constant holomorphic map such that $f(0) \neq 0, \infty, a_j$ for all j and $f'(0) \neq 0$. Let r_1 be a number ≥ 1 such that $T_f(r_1) \geq e$. There are constants $b = b(f, a_1, \ldots, a_q)$ and B_q (depending on q), such that for all $r \geq r_1$ outside a set of measure $\leq 2b_0(\psi)$ and all $b_1 \geq b_1(T_f)$ we have*

$$(q-2)T_f(r) - \sum N(f, a_j, r) + N_{f,\mathrm{Ram}}(r) \leq \frac{1}{2}S(B_q T_f^2, b_1, \psi, r) + b.$$

We can take $B_q = 12q^2 + q^3 \log 4$, for instance.

Before proving the theorem, we make some remarks on the error term. We pick ψ to grow slowly. Then 2 cancels $1/2$, and the dominant term in the error term is

$$\log T_f(r),$$

while the remaining terms have lower orders of magnitude, which can be made arbitrarily small, compatible with the convergence of the integral for ψ.

The rest of this section is devoted to the proof.

For $w, a \in \mathbf{C}$ recall that

$$\|w, a\|^2 = \frac{|w - a|^2}{(1 + |w|^2)(1 + |a|^2)} \leq 1,$$

and we make the similar definition if w or $a = \infty$, to get essentially a distance between two points on $\mathbf{P}^1$.

Remark. (Product into sum) *Let:*

$$s = \frac{1}{3} \min_{i \neq j} \|a_i, a_j\| \quad and \quad b_3 = b_3(a_1, ..., a_q) = \frac{1}{s^{2(q-1)}}$$

Then for all $w \in \mathbf{P}^1$ and all λ with $0 < \lambda < 1$ we have

$$\prod_j \|w, a_j\|^{-2(1-\lambda)} \leq b_3 \sum_j \|w, a_j\|^{-2(1-\lambda)}.$$

Indeed, for our choice of s, for all $w \in \mathbf{P}^1$ there exists at most one index j_0 such that $\|w, a_{j_0}\| \leq s$. Note that b_3 is independent of λ, again because the maximum value is reached when $\lambda = 0$.

Let Λ be a decreasing function of r with $0 < \Lambda < 1$. We allow Λ being constant. Following Ahlfors, Stoll [St], and Wong [Wo], we define the **Ahlfors-Wong function**

$$\gamma_\Lambda = \prod_j \|f, a_j\|^{-2(1-\Lambda)} \gamma_f.$$

We view a function of z as a function of (r, θ) so $\Lambda(z) = \Lambda(r)$. We have

$$-\log \gamma_\Lambda(0) \leq -\log \gamma_f(0) = b_2(f).$$

This is immediate from the fact that $\|f, a_j\| \leq 1$.

We define

$$\alpha_\Lambda = \sum_j \|f, a_j\|^{-2(1-\Lambda)} \gamma_f.$$

Then by the remark, we have the inequality

Lemma 4.3. $\gamma_\Lambda \leq b_3 \alpha_\Lambda$.

We now come to the main part of the proof.

Our first step is to estimate the integral which comes up in the Green-Jensen formula.

Proposition 4.4. *Let*

$$\Lambda(r) = \begin{cases} 1/qT_f(r) & \text{for} \quad r \geq r_1 \\ \text{constant} & \text{for} \quad r \leq r_1 \end{cases}.$$

Then

$$\frac{1}{2} \log \int_0^{2\pi} \gamma_\Lambda(re^{i\theta}) \frac{d\theta}{2\pi} \leq \frac{1}{2} S(B_q T_f^2, b_1, \psi, r) + \frac{1}{2} \log b_3.$$

for $r \geq r_1$ outside a set of measure $\leq 2b_0(\psi)$.

Let us postpone the proof of Proposition 4.4, and see how the proposition implies the theorem.

Note that for each a_j, we have

$$0 \leq \|f, a_j\|^2 = \frac{|f - a_j|^2}{(1 + |f|^2)(1 + |a_j|^2)}$$

$$= \frac{|f_1 - a f_0|^2}{(|f_0|^2 + |f_1|^2)(1 + |a_j|^2)}.$$

The function $f_1 - a f_0$ is entire and we shall count its zeros, which are precisely the points where $f = a$.

Suppose now that λ is constant. Note that for g holomorphic,

$$\log |g|^{2(1-\lambda)} = (1 - \lambda) \log |g|^2.$$

When g is meromorphic we can apply the Green-Jensen formula to $|g|^{2(1-\lambda)}$ by the homomorphic property of the log, and to a product of such factors. By Corollary 2.5, we have

$$(1-\lambda)T_f(r) - (1-\lambda)N(f, a_j, r) = -\int_0^r \frac{dt}{t}\int_{\mathbf{D}(t)} dd^c[\log \|f, a_j\|^{2(1-\lambda)}].$$

Combining this with the terms which arise from Theorem 4.1, and using the Green-Jensen formula, we get:

$$q(1-\lambda)T_f(r) - (1-\lambda)\sum N(f, a_j, r) + N_{f,\mathrm{Ram}}(r) - 2T_f(r)$$

$$= \int_0^r \frac{dt}{t}\int_{\mathbf{D}(t)} dd^c[\log \gamma_\lambda]$$

$$= \frac{1}{2}\int_0^{2\pi} \log \gamma_\lambda(re^{i\theta})\frac{d\theta}{2\pi} - \frac{1}{2}\log \gamma_\lambda(0).$$

From here on, we do not need to assume Λ *constant*, and we obtain the further inequality for arbitrary Λ:

$$\frac{1}{2}\int_0^{2\pi} \log \gamma_\Lambda(re^{i\theta})\frac{d\theta}{2\pi} - \frac{1}{2}\log \gamma_\Lambda(0) \le \frac{1}{2}\log \int_0^{2\pi}\gamma_\Lambda(re^{i\theta})\frac{d\theta}{2\pi} - \frac{1}{2}\log \gamma_f(0).$$

Let Λ be as in Proposition 4.4, that is $\Lambda = 1/qT_f$ for $r \ge r_1$. Then first we use that proposition to estimate the last integral; and second, putting $\lambda = \Lambda(r)$ we see that the bad term $-q\Lambda(r)T_f(r)$ now becomes equal to -1. Furthermore, the factor $(1-\lambda)$ before $\sum N(f, a_j, r)$ can be replaced by 1 since this sum occurs with a minus sign in front. Hence we have proved Theorem 4.2, with the constant

$$\boxed{b(f, a_1, \ldots, a_q) = \frac{1}{2}\log b_3 - \frac{1}{2}\log \gamma_f(0) + 1.}$$

Proposition 4.4 will be proved in the next section.

Remarks: With an error term $O(\log r + \log T_f(r))$ (outside an exceptional set), the theorem is due to Nevanlinna, and with the good error term, it is due to P.M. Wong [**Wo**], except for the use of the general function ψ which I suggested. In [**La 8**] I did not see how to prove my conjecture that the error term should be $(1+\varepsilon)\log T_f(r)$. As in Stoll [**St**], formula 11.7, Wong used a method of Ahlfors, namely the variable function $\Lambda(r)$. Ahlfors did not pay attention to constant factors, and used $\Lambda(r) = 1/T(r)$. Wong uses $\Lambda(r) = 1/qT(r)$, thereby getting rid of extraneous terms by this method, so that the error term is independent of $a_1, \ldots, a_q$ except in the additive constant $b(f, a_1, \ldots, a_q)$.

Ahlfors' method had been partly used by Chern [**Ch**]. However, Chern improperly reproduced this method by taking only a constant exponent λ, so that his argument is erroneous. When he takes the limit as $\lambda \to 0$, the exceptional set depends on λ, and may a priori enlarge so that it covers all the real numbers, and so Chern's argument, apparently shortcutting parts of Ahlfors, fails. The trick of Lemma 4.3, used by Wong to estimate products by sums, also comes from Ahlfors, and is a key step for getting rid of extraneous terms. I did not know this trick in [**La 8**].

Ahlfors' paper mixed the combinatorial tricks of handling the function Λ and the products into sums argument with an entirely different problem, which was to describe the differential geometry of derived curves in the context of Nevanlinna theory. Thus confusion and some ignorance about this particular combinatorial aspect of his proof was rather widespread for a long time, since nobody noticed the difficulty in Chern's paper until I pointed it out in [**La 8**]. For instance, Chern's incorrect presentation of Ahlfors' proof is reproduced in [**Gr**] p. 81.

By taking the limit illegitimately, Chern actually obtained an error term of the form

$$(\frac{1}{2} + \varepsilon)\log T_f(r),$$

just as in Theorem 4.1, estimating the ramification divisor. My guess that this is false in general, was confirmed by examples of Ye which will be found in the appendix. Two problems are involved here:

(a) In general, is $(1 + \epsilon) \log T_f(r)$ essentially the best possible error term?

(b) For each specific classical function, what is the best possible error term?

I expect that there exist examples such that $\log T_f(r)$ (asymptotically, of course) cannot be an error term (i.e. without the epsilon) possibly by constructing Weierstrass products. More generally, as in the metric Khintchine theory of the real numbers, is the error term with the functions ψ having convergent integral the best possible error term for "almost all" holomorphic functions, in a suitable sense of "almost all"? It is also a theorem of Khintchine that if φ is a positive function such that the series

$$\sum_{q=1}^{\infty} \varphi(q)$$

diverges, then for almost all real numbers α, there exist infinitely many solutions to the inequality

$$|\alpha - \frac{p}{q}| < \frac{\varphi(q)}{q}.$$

For a proof see Khintchine's book [Kh]. What is the analogue of this theorem in Nevanlinna theory? Using the type function, I have shown how to give an asymptotic estimate for the number of solutions of such an inequality with an error term involving the type function, [La 2] and [La 3] Chapter II, §3. What is the analogue of these asymptotics in Nevanlinna theory?

It is known and easy to show that the error term in Nevanlinna theory for the function e^z is $O(1)$. In number theory it is easy to show that the quadratic numbers have bounded type. It is conjectured that the only algebraic real numbers of bounded type are quadratic. Is there a similar phenomenon in Nevanlinna theory? Also there are several characterizations for numbers of bounded type, see [**La 3**] Chapter II, §2. What are the analogues in nevanlinna theory?

For the classical functions, one would have to use their special properties, individually for $\wp, \theta, \Gamma, \zeta, J$ and their variations, to determine their "type", i.e. their best possible error term in each special case. Thus I would define the **type** of a function f to be a function ψ such that the error term has the form

$$\log \psi(T_f) + O(1).$$

Finally, we observe that without much change one could formulate and prove a second main theorem when the image space is a compact Riemann surface. This will be a special case of the theorem in Chapter II. Of course, there exist holomorphic maps $f \colon \mathbf{C} \to X$ into such a surface only in case of genus 0 or 1. However, the error term is given in such a way that it applies to a map of a disc $f \colon \mathbf{D}(R) \to X$, and can be used in general to give a bound on the radius of the disc since the constants entering into the error term are given entirely as invariants of X, in the case of higher genus. This is in line with the classical Landau-Schottky theorem. The same remark applies to all the main theorems since the error term is given in explicit form each time.

I, §5. AN ESTIMATE FOR THE HEIGHT TRANSFORM

We shall prove Proposition 4.4. At first, we can work with an arbitrary Λ, not necessarily the one we selected specifically in that proposition. By Lemma 4.3 and the calculus Lemma 3.4, we find

$$\frac{1}{2} \log \int_0^{2\pi} \gamma_\Lambda(r,\theta) \frac{d\theta}{2\pi} \leq \frac{1}{2} \log \int_0^{2\pi} \alpha_\Lambda(r,\theta) \frac{d\theta}{2\pi} + \frac{1}{2} \log b_3$$

$$\leq \frac{1}{2} S(F_{\alpha_\Lambda}, \psi, r) + \frac{1}{2} \log b_3$$

for $r \geq r_1(F_{\alpha_\Lambda})$ outside a set of measure $\leq 2b_0(\psi)$.

But since $\alpha_\Lambda \geq \gamma_f$ we get $F_{\alpha_\Lambda} \geq T_f$, whence

$$r_1(F_{\alpha_\Lambda}) \leq r_1(T_f) = r_1, \qquad \text{where } T_f(r_1) \geq e$$

and we can use $r \geq r_1$ independently of Λ, as in Theorem 2.1. Furthermore the same inequality $\alpha_\Lambda \geq \gamma_f$ implies $F'_{\alpha_\Lambda} \geq T'_f$, and therefore

$$b_1(F_{\alpha_\Lambda}) \leq b_1(T_f) = b_1$$

again independently of Λ. Then we have the inequality

$$S(F_{\alpha_\Lambda}, \psi, r) \leq S(F_{\alpha_\Lambda}, b_1, \psi, r) \text{ for } r \geq r_1.$$

To estimate $S(F_{\alpha_\Lambda}, \psi, r)$ completely in terms of T_f, we shall prove:

Lemma 5.1. *We have*

$$\dot{F}_{\alpha_\Lambda} \leq \frac{q \log 4}{\Lambda^2} + \frac{12 q T_f}{\Lambda}.$$

This lemma amounts to a curvature computation, and will be proved in the next section. Now letting $\Lambda = 1/qT_f$ concludes the proof of Proposition 4.4.

Since α_Λ is additive in the a_j, it will suffice to deal with a single a_j. We restate the result to be proved above.

Proposition 5.2. *Let $a \in \mathbf{P}^1$. Let Λ be a decreasing function of r with $0 < \Lambda(r) < 1$. Define*

$$\alpha_\Lambda = \|f, a\|^{-2(1-\Lambda)} \gamma_f.$$

Then

$$\dot{F}_{\alpha_\Lambda} \leq \frac{\log 4}{\Lambda^2} + \frac{12}{\Lambda} T_f(r).$$

43

For the proof I follow P.M. Wong [**Wo**]. We shall need two lemmas.

Lemma 5.3. *Let λ be a real number, $0 < \lambda < 1$. Then*

$$\lambda^2 \|f, a\|^{-2(1-\lambda)} \gamma_f \Phi \leq 4 dd^c \log(1 + \|f, a\|^{2\lambda}) + 12\lambda \gamma_f \Phi.$$

Assuming this lemma for the moment, and putting

$$\alpha_\lambda = \|f, a\|^{-2(1-\lambda)} \gamma_f,$$

we get

$$F_{\alpha_\lambda}(r) = \int_0^r \frac{dt}{t} \int_{D(t)} \|f, a\|^{-2(1-\lambda)} \gamma_f \Phi$$

$$\leq \frac{4}{\lambda^2} \int_0^r \frac{dt}{t} \int_{D(t)} dd^c \log(1 + \|f, a\|^{2\lambda}) + \frac{12}{\lambda} T_f(r).$$

By Proposition 2.6, the Green-Jensen formula is valid in the present instance in the form we need it, that is:

Let f be meromorphic. Then

$$\int_0^r \frac{dt}{t} dd^c \log(1 + \|f, a\|^{2\lambda})$$

$$= \frac{1}{2} \int_0^{2\pi} \log(1 + \|f(re^{i\theta}), a\|^{2\lambda}) \frac{d\theta}{2\pi} - \frac{1}{2} \log(1 + \|f(0), a\|^{2\lambda}).$$

For any w we have $\|w, a\| \leq 1$. It follows that

$$F_{\alpha_\lambda}(r) \leq \frac{\log 4}{\lambda^2} + \frac{12}{\lambda} T_f(r).$$

Since Λ is a decreasing function and $0 < \Lambda(r) < 1$, for a given value of r we have

$$\|f(\zeta), a\|^{-2(1-\Lambda(|\zeta|))} \leq \|f(\zeta), a\|^{-2(1-\Lambda(r))}$$

for all $|\zeta| \leq r$. Take $\lambda = \Lambda(r)$. Then

$$F_{\alpha_\Lambda}(r) \leq F_{\alpha_\lambda}(r)$$

thus concluding the proof of Proposition 5.2.

We now come to the proof of Lemma 5.3. We shall need:

Lemma 5.4. *Let λ be a real number > 0. Then*

$$dd^c\|f, a\|^{2\lambda} = \{\lambda^2\|f, a\|^{-2(1-\lambda)} - \lambda(\lambda + 1)\|f, a\|^{2\lambda}\}\gamma_f\Phi.$$

This is done by brute force. To organize such a computation, one may wish to use the following additional formulas, which are also useful in another context. First, using the definition of $\|f, a\|^2$ as a product, and the homomorphic property of the operator $dd^c \log$, transforming multiplication into addition, and killing $|f - a|^2$, we see that

5.4.1 $$dd^c \log \|f, a\|^{2\lambda} = -\lambda\gamma_f\Phi.$$

Second we have

5.4.2 $$d\|f, a\|^2 \wedge d^c\|f, a\|^2 = \|f, a\|^2(1 - \|f, a\|^2)\gamma_f\Phi.$$

To prove this formula, first note that up to a suitable constant, we may use ∂ and $\bar{\partial}$ instead of d and d^c on the left hand side. We use the definition

$$\|f, a\|^2 = \frac{(f - a)(\bar{f} - \bar{a})}{(1 + f\bar{f})(1 + a\bar{a})}.$$

45

Then we take $\partial\|f,a\|^2$ by using the rule for the derivative of a quotient, keeping in mind that $\partial\bar f = 0$. With respect to the operator ∂, note that the factor $(\bar f - \bar a)$ behaves like a constant. We get:

$$\partial\|f,a\|^2 = \frac{\bar f - \bar a}{(1+a\bar a)}\frac{1}{(1+f\bar f)^2}[(1+f\bar f)\partial f - (f-a)\bar f \partial f]$$

$$= \frac{\bar f - \bar a}{(1+a\bar a)}\frac{1}{(1+f\bar f)^2}(1+a\bar f)\partial f.$$

Putting a complex conjugate over both sides yields $\bar\partial\|f,a\|^2$. Wedging yields

$$\partial\|f,a\|^2 \wedge \bar\partial\|f,a\|^2 = \frac{|f-a|^2}{(1+a\bar a)^2(1+f\bar f)^4}(1+a\bar f + \bar a f + a\bar a f\bar f)\partial f \wedge \bar\partial\bar f.$$

Using the definition of $\|f,a\|^2$ and substituting in the right hand side of 5.4.2 one finds the corresponding expression which proves 5.4.2.

Third, for any function u we have formulas which are verified directly from the fact that d and d^c are derivations:

dd^{c}1 $\qquad u^2 dd^c \log u = u dd^c u - du \wedge d^c u$

dd^{c}2 $\qquad dd^c \log(1+u) = \dfrac{\cdot dd^c u}{1+u} - \dfrac{du \wedge d^c u}{(1+u)^2}$

dd^{c}3 $\qquad\qquad = \dfrac{dd^c u}{(1+u)^2} + \dfrac{u^2 dd^c \log u}{(1+u)^2}$ [using **dd^{c}1**]

dd^{c}4 $\qquad\qquad = \dfrac{1}{u(1+u)^2} du \wedge d^c u + \dfrac{u}{1+u} dd^c \log u.$

Now to prove Lemma 5.4, let $u = \|f,a\|^2$. Then the left hand side in Lemma 5.4 is $dd^c u^\lambda$, and we have

$$dd^c u^\lambda = d(\lambda u^{\lambda-1} d^c u) = \lambda(\lambda-1)u^{\lambda-2}du \wedge d^c u + \lambda u^{\lambda-1}dd^c u.$$

Note that 5.4.2 gives $du \wedge d^c u$ in terms of $\gamma_f \Phi$, and formula **dd^{c}1** gives

$$dd^c u = u dd^c \log u + u^{-1} du \wedge d^c u.$$

46

Therefore again we can use 5.4.2 combined with 5.4.1 to get $dd^c u$ in terms of $\gamma_f \Phi$. If one combines all the factors of $\gamma_f \Phi$, one finds precisely the factor on the right hand side of Lemma 5.4.

We now conclude the proof of Lemma 5.3, that is for $0 < \lambda < 1$ we prove the inequality

$$dd^c \log(1 + ||f,a||^{2\lambda}) \geq \frac{\lambda^2}{4}||f,a||^{-2(1-\lambda)}\gamma_f \Phi - 3\lambda\gamma_f \Phi.$$

Let $u = ||f,a||^{2\lambda}$. Then using Lemma 5.4 and 5.4.1 we get

$$dd^c \log(1 + ||f,a||^{2\lambda}) = \frac{dd^c ||f,a||^{2\lambda}}{(1 + ||f,a||^{2\lambda})^2} + \frac{||f,a||^{4\lambda} dd^c \log ||f,a||^{2\lambda}}{(1 + ||f,a||^{2\lambda})^2}$$

$$\geq \{\frac{1}{4}\lambda^2 ||f,a||^{-2(1-\lambda)} - \lambda(\lambda+1)||f,a||^{2\lambda}\}\gamma_f \Phi - \lambda\gamma_f \Phi$$

using the fact that $||f,a|| \leq 1$. Replacing $\lambda + 1$ by 2 and combining the last two terms concludes the proof of Lemma 5.3, and therefore also of Proposition 5.2.

Note: The result here in its precise form is due to P.M. Wong [**Wo**], and replaces a curvature computation. A weaker version appears in Stoll [**St**], as an "Ahlfors Estimate", Theorem 10.3. The introduction of curvature in Nevanlinna theory is due to F. Nevanlinna (cf. [**Ne**] Chapter IX, §4), but the way it is carried out here, using the singular form with variable Λ is closer to Ahlfors [**Ah**]. Indeed, the occurrence of T_f^2 in the error term is already in Ahlfors, pp. 26-27 in a similar curvature computation, although as usual, Ahlfors does not keep track of constants. It is the presence of T_f^2 in Theorem 4.2 and Proposition 4.4 combined with the factor $1/2$ in front which gave rise to the proof of my conjecture that the error term should be of the form $(1 + \varepsilon)\log T_f(r)$. As long as one was not looking for the best possible error term, no great significance was attached to this particular structure of the proof, and the significance of T_f^2 went unnoticed.

I, §6. VARIATIONS AND APPLICATIONS. THE LEMMA ON THE LOGARITHMIC DERIVATIVE

We start by giving a proof of Nevanlinna's lemma on the logarithmic derivative. The basic differential geometric pattern was given by Nevanlinna [Ne] p. 259. I did not quite get the desired error term in [La 8], but I am now able to get it by using Wong's idea, with the Ahlfors-Wong function Λ.

We write m_f for $m_{f,\infty}$ and similarly for T_f.

Theorem 6.1 (Lemma on the logarithmic derivative). *Let f be a non-constant meromorphic function such that $f(0) \neq 0, \infty$ and $f'(0) \neq 0$. Then*

$$m_{f'/f}(r) \leq \frac{1}{2} S(B_2 T_f^2, \psi, b_1, r) - \log \frac{|f'(0)|^2}{(1+|f(0)|^2)^2} + 1 + \frac{1}{2} \log b_3(0, \infty)$$

for $r \geq r_1$ outside a set of measure $\leq 2b_0(\psi)$.

Proof: Let $\Lambda = 1/2T_f$ for $r \geq r_1$ and constant for $r \leq r_1$, as in Proposition 4.4. We let $D = (0) + (\infty)$, so $q = 2$. Then directly from the definitions,

$$\gamma_\Lambda = |f'/f|^2 h^\Lambda \qquad \text{where} \qquad h = \|f, 0\|^2 \|f, \infty\|^2.$$

Let

$$u = |f'/f|^2 \qquad \text{and} \qquad v = h^\Lambda, \quad \text{so} \quad 0 \leq v \leq 1.$$

Denote by $S(r)$ the circle of radius r, and let $\sigma = d\theta/2\pi$. Then

$$m_{f'/f}(r) = \frac{1}{2} \int_{S(r)} (\log^+ u)\sigma$$

$$= \frac{1}{2} \int_{S(r)} (\log^+ u + \log v)\sigma - \frac{1}{2} \int_{S(r)} (\log v)\sigma$$

$$= \frac{1}{2} \int_{S(r)} (\log e^{\log^+ u + \log v})\sigma - \frac{1}{2}\Lambda(r) \int_{S(r)} (\log h)\sigma$$

$$\leq \frac{1}{2} \log \int_{S(r)} uv\sigma + 1 + \Lambda(r)m_{f,(0)+(\infty)}(r).$$

Indeed, for this last inequality, we pull the first log out of the integral. Then we use that if $0 \leq v \leq 1$ then

$$e^{\log^+ u + \log v} \leq uv + 1 = \gamma_\Lambda + 1,$$

which gives us the first term by Proposition 4.4. As to the second term,

$$m_{f,(0)+(\infty)} \leq 2T_f - \log \frac{|f'(0)|^2}{(1 + |f(0)|^2)^2}.$$

Hence our choice of $\Lambda(r)$ gives the desired bound.

The lemma on the logarithmic derivative was essentially a variation on the Second Main Theorem. Next, we give an actual application of this theorem due to Nevanlinna, and bounding the number of totally ramified values.

Given a point $a \in \mathbf{P}^1$ we consider those elements $z \in \mathbf{C}$ such that $f(z) = a$. We say that f is **totally ramified over** a if for every such element z we have

$$\mathrm{ord}_z(f - a) > 1.$$

Theorem 6.2. *Let f be a non-constant meromorphic function. Then f has at most four totally ramified values.*

49

Proof: Let $a_1, \ldots, a_q$ be distinct totally ramified values. We have to prove that $q \leq 4$. Write

$$n_f(a, r) = n_f^*(a, r) + n_{f,\text{Ram}}(a, r)$$

where $n_f^*(a, r)$ is the number of $z \in \mathbf{D}(r)$ such that $f(z) = a$, counted with multiplicity 1, and $n_{f,\text{Ram}}(a, r)$ is the number counted with the ramification index, i.e. if f is ramified of order e at z_0 then z_0 is counted with multiplicity $e-1$, then we get the corresponding counting functions

$$N_f^*(a, r) + \int_0^r n_f^*(a, t) \frac{dt}{t} + n_f^*(a, 0) \log r,$$

so

$$N_f(a, r) = N_f^*(a, r) + N_{f,\text{Ram}}(a, r).$$

By the Second Main Theorem,

$$\sum_{j=1}^q m_f(a_j, r) + N_{f,\text{Ram}}(r) - 2T_f(r) \leq o_{\text{exc}}(T_f(r)).$$

Here I use an extension o_{exc} of the usual symbol o. Namely, I define

$$\alpha = o_{\text{exc}}(\beta)$$

to mean that there exists a set E of finite measure such that

$$\lim_{\substack{r \to \infty \\ r \notin E}} \frac{\alpha(r)}{\beta(r)} = 0.$$

Using our expression for $N_{f,\text{Ram}}$, we find by the First Main Theorem:

$$(q - 2)T_f(r) \leq \sum_{j=1}^q N_f^*(a_j, r) + o_{\text{exc}}(T_f(r))$$

$$\leq \sum_{j=1}^q \frac{1}{2} N_f(a_j, r) + o_{\text{exc}}(T_f(r))$$

$$\leq \frac{q}{2} T_f(r) + o_{\text{exc}}(T_f(r)).$$

Hence
$$\frac{q}{2}T_f(r) \leq 2T_f(r) + o_{\mathrm{exc}}(T_f(r)),$$
which proves $q \leq 4$, and concludes the proof of the theorem.

Note that the Weierstrass $\wp$-function has exactly 4 totally ramified values, so 4 is best possible in general.

Appendix. On Nevanlinna's Error Term

by Zhuan Ye

This note is a part of an ongoing more detailed study. In Theorem 4.2 of Chapter I of these notes, Serge Lang presents the explicit estimate of Pit-Mann Wong [Wo] conjectured by Lang that

(1)
$$\sum_{\nu=1}^{q} m(r, a_\nu) - 2T(r, f) + N_{\text{Ram}}(r, f) \leq \log T(r, f) + \text{ lower-order terms}$$

outside an exceptional set as $r \to \infty$. We show here that this estimate is essentially best possible, in the sense that "1" on the right side of (1) cannot be replaced by a smaller number. We have the following

Theorem. *Given $\varepsilon > 0$, there exists an entire function f of finite order and a finite set $\{a_\nu\}$ such that*

(2) $$\sum m(r, a_\nu) - 2T(r, f) + N_{\text{Ram}}(r, f) > (1 - \epsilon) \log T(r, f),$$

for all large r.

Proof: We consider the function

$$f(z) = \int_0^z e^{-t^q} dt$$

where $q \geq 2$ is a integer, $a_\nu = e^{2\pi i \nu / q} \int_0^\infty e^{-t^q} dt$ and examine the computation of [Ne, VI §2.3].
If $|\arg z - 2\pi \nu q^{-1}| \leq \frac{1}{2} \pi q^{-1}$, we see that

$$f(z) - a_\nu = -\int_z^\infty e^{-t^q} dt = -\frac{e^{-z^q}}{q z^{q-1}} + \frac{q-1}{q} \int_z^\infty \frac{e^{-t^q}}{t^q} dt$$

$$= -\frac{e^{-z^q}}{q z^{q-1}}(1 + o(1)), \qquad (1 \leq \nu \leq q).$$

53

Thus, when $|\arg z - 2\pi\nu q^{-1}| \le \frac{1}{2}\pi q^{-1}$, we find that

$$\log|f(re^{i\theta}) - a_\nu| = -(r^q \cos q\theta + (q-1)\log r + O(1)).$$

Since the a_ν are distinct, this shows that

$$m(r, a_\nu, f) = \frac{1}{\pi q}r^q + \frac{q-1}{2q}\log r + O(1), \qquad (1 \le \nu \le q).$$

The computation of $T(r, f)$ is similar. In

$$|\arg z - (2\nu - 1)\pi q^{-1}| < \frac{1}{2}\pi q^{-1},$$

we find that

$$f(z) = \frac{e^{-z^q}}{qz^{q-1}}(1 + o(1)) + O(1),$$

and hence, on summing over these q sectors,

$$T(r, f) = m(r, f) = \frac{1}{\pi}r^q - \frac{q-1}{2}\log r + O(1).$$

Thus, $\log T(r, f) = (q - 1)\log r + O(1)$. In particular, $f(z)$ has order q, mean type. Obviously $N_{\text{Ram}}(r, f) \equiv 0$, and hence

$$\sum_{\nu=1}^{q} m(r, f, a_\nu) + m(r, f) - 2T(r, f) + N_{\text{Ram}}(r, f)$$

$$= (q - 1)\log r + O(1) = \frac{q-1}{q}\log T(r, f) + O(1).$$

If we choose q so large that $q\epsilon > 1$, we have shown (2).

This example may be typical. For example, if we consider $g(z) = e^{z^p}$, slightly easier computations show that $m(r, g, 0) = m(r, g) = T(r, g) = \frac{1}{\pi}r^p$, so that

$$\log T(r, g) = p \, \log r + O(1).$$

Finally, the only ramification is at the origin, where $f(z)$ has a zero of multiplicity $p - 1$. Thus

$$m(r, g, 0) + m(r, g) + N_{\text{Ram}}(r, g) - 2T(r, g)$$
$$\equiv N_{\text{Ram}}(r, g) = (p - 1) \log r + O(1)$$
$$= \frac{p - 1}{p} \log T(r, g) + o(1).$$

By choosing p large, we again obtain (2).

In view of the ubiquity of such examples, it seems interesting to find the exact bound in (1).

Finally, I thank D. Drasin for presenting this problem to me.

CHAPTER II

EQUIDIMENSIONAL HIGHER DIMENSIONAL THEORY

Here following Carlson-Griffiths [C-G] we consider the higher dimensional situation which is closest to that of dimension one, namely a mapping

$$f \colon \mathbf{C}^n \to X$$

where X is a compact complex manifold of dimension n. Then the theorems and proofs are completely parallel, with no additional difficulty except that which comes from the extra formal manipulation of higher dimensions. But the theory is in good shape, in that once one has understood the basic principles and mechanisms which must be introduced to deal with the higher dimensional case, then the translation from dimension one becomes essentially automatic. These mechanisms are essentially fundamental to complex analysis and complex differential geometry, especially concerning holomorphic curvature, and are of interest independently of Nevanlinna theory.

Carlson-Griffiths used a singular volume form more or less generalizing such use by F. Nevanlinna. Here we follow mostly Ahlfors and P.M. Wong [Wo], who uses a more efficient singular form on $\mathbf{C}^n$ instead of X.

II, §1. THE CHERN AND RICCI FORMS

We begin by fixing some notation and terminology.

We recall the notion of a divisor, both on X and on $\mathbf{C}^n$, so let Y be a complex manifold, not necessarily compact. Consider pairs (U, φ)

consisting of an open set U and a meromorphic function φ. We say that two pairs (U, φ) and (V, ψ) are **equivalent** if $\varphi\psi^{-1}$ is holomorphic invertible on $U \cap V$, so $\varphi^{-1}\psi$ is also holomorphic. By a (Cartier) **divisor** on Y we mean a maximal family of equivalent pairs $\{(U_i, \varphi_i)\}$ such that the open sets U_i cover Y. If one is given a family of such pairs which is not maximal, but such that the U_i cover Y then we say that this family **represents** the divisor. If (U, φ) is a pair in the family, we say that this pair **represents** the divisor on U, or that φ represents the divisor on U. Thus a divisor is defined by local conditions. If all φ_i are holomorphic, we say that the divisor is **effective**. On the whole we assume that the reader is acquainted with elementary properties of divisors, e.g. that they form a group, the existence of pull backs, etc. It is a basic fact that on $\mathbf{C}^n$, given a divisor Z, there always exists a meromorphic function g on $\mathbf{C}^n$ which represents the divisor globally, namely $(\mathbf{C}^n, g)$ is a representative pair for the divisor. Furthermore, on $\mathbf{C}^n$ every meromorphic function is the quotient of two holomorphic functions. See Gunning-Rossi [**GuR**] theorems III-N4, III-K5 and III-K6. Of course such properties cannot be true on compact manifolds: they have to do with so-called Stein manifolds. However if

$$ f : \mathbf{C}^n \to X $$

is holomorphic and D is a divisor on X such that $f(\mathbf{C}^n)$ is not contained in D, then the pull back $f^{-1}(D)$ is a divisor on $\mathbf{C}^n$, which can therefore be represented by a meromorphic function on $\mathbf{C}^n$.

Given a divisor defined on an open set U by a function φ. Let O be the ring of holomorphic functions at a point of U. Then O has unique factorization, and we can factorize

$$ \varphi = \varphi_1^{m_1} \cdots \varphi_r^{m_r} $$

where $\varphi_1, \ldots, \varphi_r$ are irreducible elements. If we replace φ by the product

$$ \varphi_1 \cdots \varphi_r $$

and do this for every pair (U, φ), then we obtain another divisor, called the **reduced divisor**. If $r = 1$ in the above expression, and $m_1 = 1$, then we say that the divisor is **irreducible locally**. Globally, every divisor on a compact manifold can be written as a sum of irreducible ones.

Let X be a complex manifold of dimension n, and let L be a holomorphic line bundle over X. As we deal only with holomorphic bundles, unless otherwise specified, we shall omit the word holomorphic to qualify them. Let $\{U_i\}$ be an open covering of X such that $L \mid U_i$ has a trivialization (holomorphic, according to our convention)

$$\varphi_i \colon L \mid U_i \to U_i \times \mathbf{C}.$$

Then

$$\varphi_{ij} = \varphi_i \circ \varphi_j^{-1} \colon (U_i \cap U_j) \times \mathbf{C} \to (U_i \cap U_j) \times \mathbf{C}$$

is an isomorphism, given by a holomorphic map

$$g_{ij} \colon U_i \cap U_j \to \mathbf{C}^* = GL_1(\mathbf{C})$$

such that

$$\varphi_{ij}(x, z) = (x, g_{ij}(x)z).$$

Let s be a holomorphic section of L over X. Then s is represented by a holomorphic map

$$s_i \colon U_i \to \mathbf{C},$$

satisfying

$$s_i = g_{ij} s_j.$$

Suppose a covering family $\{(U_i, \varphi_{ij})\}$ represents L as above. Suppose given for each i a function (smooth)

$$\rho_i \colon U_i \to \mathbf{R}_{>0}$$

such that on $U_i \cap U_j$ we have

$$\rho_i = |g_{ij}|^2 \rho_j.$$

Then we say that the family of triplets $\{(U_i, \varphi_{ij}, \rho_i)\}$ **represents a metric.** It is clear how to define compatible families, or compatible triples in this context, and the metric itself is an equivalence class of such covering triples, or is the maximal family of compatible triples. We could also write a representative family as $\{(U_i, \varphi_i, \rho_i)\}$ using the isomorphisms φ_i instead of the transition functions φ_{ij}.

If s is a section of L over some open set U_i containing a point P, then we define

$$|s(P)|^2 = \frac{|s_i(P)|^2}{\rho_i(P)}.$$

The value on the right-hand side is independent of the choice of U_i, as one sees at once from the transformation law.

Instead of using indices i, if L is trivial over an open set U, so

$$L_U \approx U \times \mathbf{C},$$

we write $s_U \colon U \to \mathbf{C}$ for the map representing a section $s \colon U \to L$ over U, and then we also write

$$|s|^2 = |s_U|^2 / \rho_U.$$

A metric as above will be denoted by ρ, for instance. Since at each point L is one-dimensional, we see that the metric is determined by a hermitian product in a trivial way. In particular, we obtain line bundles by considering the tangent or cotangent bundle of a one-dimensional complex manifold, usually called a **Riemann surface.**

We shall now associate some differential forms to a metric on a line bundle. First we review some terminology.

Let $z_1, \ldots, z_n$ be holomorphic coordinates for X over U. As usual, we have the operators ∂ and $\bar{\partial}$, where say for a function $f(z)$,

$$\partial f(z) = \sum_{k=1}^{n} \frac{\partial f}{\partial z_k} dz_k \quad \text{and} \quad \bar{\partial} f(z) = \sum_{k=1}^{n} \frac{\partial f}{\partial \bar{z}_k} d\bar{z}_k.$$

Then

$$d = \partial + \bar{\partial}.$$

The operators $\partial, \bar{\partial}$, and d extend to forms of arbitrary degree as usual, for instance

$$\bar{\partial}(f_{IJ}(z, \bar{z}) dz_I \wedge d\bar{z}_J) = \sum_{k=1}^{n} \frac{\partial f}{\partial \bar{z}_k} d\bar{z}_k \wedge dz_I \wedge d\bar{z}_J,$$

where $dz_I = dz_{i_1} \wedge \cdots \wedge dz_{i_p}$ and $d\bar{z}_J = d\bar{z}_{j_1} \wedge \cdots \wedge d\bar{z}_{j_q}$. A sum of terms

$$\omega = \sum_{\substack{|I|=p \\ |J|=q}} f_{IJ}(z, \bar{z}) dz_I \wedge d\bar{z}_J$$

is called a form of **type** (p, q). The numbers p, q do not depend on the choice of holomorphic coordinates, because if g is holomorphic, then $\bar{\partial} g = 0$ by the Cauchy-Riemann equations.

We define the operator

$$d^c = \frac{1}{4\pi \sqrt{-1}} (\partial - \bar{\partial}).$$

The advantage of such an operator is that it is a real operator. If ω is a form such that $\omega = \bar{\omega}$, then $d^c \omega$ also satisfies this property. Note that

$$dd^c = \frac{\sqrt{-1}}{2\pi} \partial \bar{\partial} = \frac{1}{2\pi \sqrt{-1}} \bar{\partial} \partial.$$

In dimension 1, and polar coordinates (r, θ) we have

$$\boxed{d^c = \frac{1}{2} r \frac{\partial}{\partial r} \otimes \frac{d\theta}{2\pi} - \frac{1}{4\pi} \frac{1}{r} \frac{\partial}{\partial \theta} \otimes dr.}$$

.

Restricting to the circle the term with dr vanishes, so

$$d^c \text{ restricted to the circle is } \frac{1}{2}r\frac{\partial}{\partial r}\frac{d\theta}{2\pi}.$$

Given a metric ρ on L we define the **Chern form** of the metric to be the unique form $c_1(\rho)$ such that on an open set U, in terms of the trivialization as above, we have

$$\boxed{c_1(\rho) \mid U = -dd^c \log |s|^2 = dd^c \log \rho_U}$$

for any holomorphic section s. The right-hand side is independent of the choice of holomorphic coordinates on U, because for any non-zero holomorphic function g (giving rise to a change of charts) we have

$$\partial\bar{\partial} \, \log |g|^2 = \partial\bar{\partial} \, \log(g\bar{g}) = 0.$$

so $dd^c \log(g\bar{g}) = 0$.

We say that a $(1,1)$-form

$$\omega = \frac{\sqrt{-1}}{2\pi} \sum h_{ij}(z)dz_i \wedge d\bar{z}_j$$

is **positive** and we write $\omega > 0$, if the matrix $h = (h_{ij})$ is hermitian positive definite for all values of z. This condition is independent of the choice of holomorphic coordinates $z_1, \ldots, z_n$.

A metric ρ is called **positive** if $c_1(\rho)$ is positive.

Example. Projective space. Let $X = \mathbf{P}^n$ be projective n-space. Let $T_0, \ldots, T_n$ be the homogeneous variables, and let U_i be the open set of points such that $T_i \neq 0$. We let

$$z_0^{(i)} = T_0/T_i, \ldots, z_n^{(i)} = T_n/T_i \qquad \text{so} \quad z_i^{(i)} = 1.$$

Then $z_j^{(i)}$ with $j \neq i$ are complex coordinates on U_i. Then there is a line bundle L, called the **hyperplane line bundle**, whose transition functions are given by

$$g_{ij} = T_j/T_i.$$

Its sheaf of sections is usually denoted by $\mathcal{O}(1)$. Say for $i = 0$, we write simply $z = (z_1, \ldots, z_n)$ where $z_j = T_j/T_0$. The **standard metric** ρ on L is defined on U_0 by the function

$$\rho(z) = 1 + \sum_{\nu=1}^{n} z_\nu \bar{z}_\nu,$$

and similarly for U_i. Then its Chern form is computed using rules from freshman calculus for the derivative of a product and quotient, to give

$$c_1(\rho) = \frac{\sqrt{-1}}{2\pi} \partial\bar{\partial} \, \log \rho(z)$$

$$= \frac{\sqrt{-1}}{2\pi} \frac{1}{\rho(z)^2} \Big(\sum_{i,j=1}^{n} h_{ij} dz_i \wedge d\bar{z}_j \Big),$$

where $h = (h_{ij})$ is the matrix $h = \rho(z)I - (\bar{z}_i z_j)$. The metric on the cotangent bundle defined by this Chern form is called the **Fubini-Study metric**.

Proposition 1.1 . *The Fubini-Study metric is positive.*

Proof: We have to show that h is positive definite (it is obviously hermitian). For any complex vector $C = {}^t(c_1, \ldots, c_n)$ we expand ${}^t\bar{C}hC$ and the Schwartz inequality immediately shows that for $C \neq 0$ we have

$${}^t\bar{C}hC > 0.$$

This proves the proposition.

The example will play no role for the rest of this chapter, but becomes useful later.

We now pass to volume forms. In $\mathbf{C}^n$ we have what we call the **euclidean form**, expressed in terms of coordinates z by

$$\boxed{\Phi(z) = \prod_{i=1}^{n} \frac{\sqrt{-1}}{2\pi} dz_i \wedge d\bar{z}_i.}$$

Except for the factor involving π and the power of 2, it is just

$$dx_1 \wedge dy_1 \wedge \cdots \wedge dx_n \wedge dy_n.$$

The product sign is to be interpreted as the alternating product, but 2-forms commute with all forms, so it is harmless to write it as the usual product sign to emphasize this commutativity.

By a **volume form** on X, we mean a form of type (n, n), which locally in terms of complex coordinates can be written as

$$\Psi(z) = h(z)\Phi(z),$$

where h is C^∞ and $h(z) > 0$ for all z. This is invariant under a change of complex coordinates, since the factor coming out in such a change is of the form $g(z)\overline{g(z)}$, where $g(z)$ is holomorphic invertible.

Thus a volume form is a metric on the **canonical bundle**

$$K_X = \overset{\text{max}}{\bigwedge} T^\vee(X),$$

which is the $\max{}(n\,\text{th})$ exterior power of the cotangent bundle.

We define the **Ricci form** of Ψ to be the Chern form of this metric, so $\mathrm{Ric}(\Psi)$ is the *real* $(1, 1)$-form given by

$$\mathrm{Ric}(\Psi) = c_1(\rho) = dd^c \log h(z) \quad \text{in terms of coordinates } z.$$

Remarks: *If C is a constant then*

$$\mathrm{Ric}(C\Psi) = \mathrm{Ric}(\Psi).$$

If u is a positive smooth function, then

$$\mathrm{Ric}(u\Psi) = \mathrm{Ric}(\Psi) + dd^c \log u.$$

Both assertions are trivial from the definition.

A 2-form commutes with all forms. By the n-th power

$$\mathrm{Ric}(\Psi)^n$$

we mean the n-th exterior power. Then $\mathrm{Ric}(\Psi)^n$ is an (n, n)-form, and in particular a max degree form on X. Since Ψ is a volume form, there is a unique function G on X such that

$$\frac{1}{n!}\mathrm{Ric}(\Psi)^n = G\Psi.$$

We may also write symbolically

$$G = \frac{1}{n!}\mathrm{Ric}(\Psi)^n/\Psi.$$

Note that G is a real-valued function. We call G the **Griffiths function** associated with the original volume form Ψ. We denote it by

$$G_\Psi \quad \text{or} \quad G(\Psi).$$

Special case *Let dim $X = 1$, and let z be a complex coordinate. Then the Griffiths function is given by*

$$G_\Psi(z) = \frac{1}{h(z)}\frac{\partial^2 \log h(z)}{\partial z \partial \bar{z}}.$$

Proof: Immediate from the definitions.

The function $-G$ in dimension 1 is classically called the **Gauss curvature**.

Remark: In the above definitions, we did not need to assume X compact. If X is compact, we shall also deal with volume forms Ψ which are defined only on the complement of a divisor in X. Then the same definitions will apply to such forms. In particular, we have the Ricci form and the Griffiths function defined outside the singularities.

II, §2. SOME FORMS ON $\mathbf{C}^n$ AND $\mathbf{P}^{n-1}(\mathbf{C})$ AND THE GREEN-JENSEN FORMULA

We shall deal constantly with the following differential forms on $\mathbf{C}^n$, letting $z = (z_1, \ldots, z_n)$ be complex coordinates:

$$\omega(z) = dd^c \log \|z\|^2$$
$$\varphi(z) = dd^c \|z\|^2$$
$$\sigma(z) = d^c \log \|z\|^2 \wedge \omega^{n-1}$$

The form φ will appear a little later, at first we deal with ω and σ. We shall in fact deal with the restriction of σ to the sphere $\mathbf{S}(r)$ for $r > 0$.

Remark: Let

$$\pi \colon \mathbf{C}^n - \{0\} \to \mathbf{P}^{n-1}$$

be the natural map representing a point of projective space by homogeneous coordinates. Let τ be the Fubini-study metric, and let $\omega_{\mathbf{P}} = c_1(\tau)$ be the Fubini-Study form. Then

$$\omega = \pi^* \omega_{\mathbf{P}}.$$

This is immediate from the definitions. The fibering $\pi \colon \mathbf{C}^n - \{0\} \to \mathbf{P}^{n-1}$ restricts to a fibering on each sphere

$$\pi \colon \mathbf{S}(r) \to \mathbf{P}^{n-1}$$

which we use for the rest of this section.

We can write the original formula for $\omega_{\mathbf{P}}$ on $U_0 = \mathbf{C}^n$ in the form

$$\omega_{\mathbf{P}} = \frac{\sqrt{-1}}{2\pi} \frac{1}{1 + \|z\|^2} \left(<dz, dz> - \frac{<dz, z> \wedge <z, dz>}{1 + \|z\|^2} \right),$$

with the notation of the hermitian product $<dz, dz> = \sum dz_k \wedge d\bar{z}_k$ for instance. The exterior powers (other than the first) of $<dz, z>$

and $< z, dz >$ are 0. Hence

$$\left(\frac{2\pi}{\sqrt{-1}}\right)^n (1+\|z\|^2)^n \, \omega_{\mathbf{P}}^n$$

$$= \left(< dz, dz >^n - \frac{n < dz, dz >^{n-1} \wedge < dz, z > \wedge < z, dz >}{1+\|z\|^2}\right).$$

But

$$< dz, dz >^n = n! \prod dz_k \wedge d\bar{z}_k \qquad \text{and}$$

$$< dz, dz >^{n-1} = (n-1)! \sum dz_1 \wedge d\bar{z}_1 \wedge \cdots \wedge \widehat{dz_k \wedge d\bar{z}_k} \wedge \cdots \wedge dz_n \wedge d\bar{z}_n,$$

so we get the formula

$$\omega_{\mathbf{P}}^n(z) = \frac{n!}{(1+\|z\|^2)^{n+1}} \Phi(z).$$

We shall use **Fubini's theorem** in a global version. In general, let

$$\pi: X \to Y$$

be a fibering of real manifolds, meaning a C^∞ map of manifolds, locally isomorphic to a product. Let $q = \dim Y$ and $\dim X = p+q$, so p is the dimension of a fiber. Let η be a p-form on X and let ω_Y be a q-form on Y. Then

$$\int_X \pi^* \omega_Y \wedge \eta = \int_{y \in Y} \left(\int_{\pi^{-1}(y)} \eta\right) \omega_Y(y),$$

under conditions of absolute convergence. For instance, if η has compact support, then the formula is valid. The proof reduces to a local statement by partitions of unity, and locally, the relation is merely Fubini's theorem that a double integral is equal to a repeated integral.

Proposition 2.1. $\displaystyle \int_{S(r)} \sigma = 1 = \int_{\mathbf{P}^{n-1}} \omega_{\mathbf{P}}^{n-1}.$

Proof: We consider the fibering

$$\pi : S(r) \to \mathbf{P}^{n-1}.$$

Let I_n be the integral on the left of the proposition, and let J_{n-1} be the integral on the right. For $\pi : S(r) \to \mathbf{P}^{n-1}$ we get

$$I_n = \int_{S(r)} \sigma = \int_{y \in \mathbf{P}^{n-1}} \left(\int_{\pi^{-1}(y)} d^c \log \|z\|^2 \right) \omega_{\mathbf{P}}^{n-1}(y)$$

$$= \int_{\pi^{-1}(y)} d^c \log \|z\|^2 \cdot J_{n-1}$$

$$= J_{n-1},$$

because the first factor is independent of $y \in \mathbf{P}^{n-1}$, and one computes directly in dimension 1, using the formula for d^c given in polar coordinates by

$$d^c = \frac{1}{2} r \frac{\partial}{\partial r} \otimes \frac{d\theta}{2\pi} \qquad \text{on the circle.}$$

Now to compute J_n, since the complement of U_0 is a hyperplane, having measure 0, we have

$$J_n = \int_{\mathbf{P}^n} \omega_{\mathbf{P}}^n = \int_{\mathbf{C}^n} \omega_{\mathbf{P}}^n = n! \int_{\mathbf{C}^n} \frac{1}{(1 + \|z\|^2)^{n+1}} \prod_{k=1}^{n} \frac{\sqrt{-1}}{2\pi} dz_k \wedge d\bar{z}_k.$$

This is now easily evaluated by induction, leaving $z_1, \ldots, z_{n-1}$ fixed and integrating with respect to z_n, using polar coordinates $z_n = re^{i\theta}$. One finds that $J_n = J_{n-1}$, whence $J_n = 1$ by a direct computation on J_1. This concludes the proof.

The next proposition generalizes the formula for d^c in higher dimension.

Proposition 2.2. *For smooth functions α, we have*

$$d^c \alpha \wedge \omega^{n-1} | S(t) = \frac{1}{2} t \frac{\partial \alpha}{\partial t} \sigma$$

68

Proof: Apply the left-hand side as a distribution, i.e., as a functional on smooth functions with compact support. Again use the fibration

$$\pi: \mathbf{S}(r) \to \mathbf{P}^{n-1}.$$

Suppose β is C^∞ on $\mathbf{S}(t)$. It suffices to verify

$$\int_{y \in \mathbf{P}^{n-1}} \left(\int_{\pi^{-1}(y)} \beta d^c \alpha \right) \omega_Y^{n-1}(y) = \int_{y \in \mathbf{P}^{n-1}} \frac{r}{2} \left(\int_{\pi^{-1}(y)} \beta \frac{\partial \alpha}{\partial r} \frac{d\theta}{2\pi} \right) \omega_Y^{n-1}(y),$$

which is true because $\pi^{-1}(y)$ is a circle, and on the circle we already know the formula for d^c. Thus the formula follows in general.

Observe that the formula extends to more general functions β by general integration theory.

We now want to extend the theorems in one variable having to do with various integrations and the height transform. We shall consider functions whose singularities are on divisors of $\mathbf{C}^n$. They will be of three types:

I. C^∞ functions, for which no problem will arise in applying Stokes' theorem, or differentiating under the integral sign.

II. Functions which have singularities on a divisor of $\mathbf{C}^n$, and which are locally of the form

$$\alpha = \log |g|^2$$

where g is holomorphic.

III. Functions locally of the form $\log(1 + (|g|^2 h)^\lambda)$ with $0 < \lambda \leq 1$, where h is > 0 and C^∞ and g is holomorphic.

We shall also consider linear combinations of these functions with constant coefficients.

69

We could call the space generated by such functions the vector space of **admissible functions**. I have not found yet a very easy characterization of functions for which the formal arguments which we shall apply are valid. For our purposes at the moment, we shall suspend the definition of an appropriate space for which the formal arguments which follow are valid.

Theorem 2.3 *For any admissible function α which is C^∞ near the origin, we have*

$$\int\limits_0^r \frac{dt}{t} \int\limits_{S(t)} d^c\alpha \wedge \omega^{n-1} = \frac{1}{2} \int\limits_{S(r)} \alpha\sigma - \frac{1}{2}\alpha(0).$$

Proof: By Proposition 2.2, and after permuting a derivative with the integral, we have

$$\int\limits_0^r \frac{dt}{t} \int\limits_{S(t)} d^c\alpha \wedge \omega^{n-1} = \int\limits_0^r \frac{dt}{t}\frac{1}{2}t\frac{\partial}{\partial t} \int\limits_{S(t)} \alpha\sigma$$

$$= \frac{1}{2} \int\limits_{S(r)} \alpha\sigma - \frac{1}{2}\alpha(0)$$

thus proving the theorem.

II, §3. STOKES' THEOREM WITH CERTAIN SINGULARITIES ON $\mathbf{C}^n$

Although we ultimately study holomorphic mappings $f: \mathbf{C}^n \to X$ into a compact complex manifold (in fact, algebraic), these are studied via pull-backs to $\mathbf{C}^n$, and basic relations for f are reduced to certain relations on $\mathbf{C}^n$. These relations involve Stokes' theorem, and we shall now develop them directly on $\mathbf{C}^n$.

A version of Stokes' theorem with singularities is given in [**La 5**] and [**La 6**]. The idea is that if the singularities are of dimension one less than the dimension of the set of regular points on the boundary of the manifold in question, then Stokes' formula is valid. The integral of a form over an analytic space is by definition the integral over the set of its regular points. Since this set may not be compact, it is of course necessary to know the absolute convergence of all integrals involved. The proof of Stokes' formula with singularities is obtained as follows. In that reference, an a priori definition of a **negligible** set on the boundary is given, such that if the singularities are negligible then Stokes' formula applies. One multiplies the given form by a C^∞ function that is 0 near the singularities and 1 slightly farther out from the singularities. One then takes a limit over such functions, depending on a parameter which tends to 0. One has to estimate the derivatives that come into the proof when taking d of the form. All details are given in the above references to carry out this idea.

Throughout this section, we let Z be a divisor on $\mathbf{C}^n$ and $0 \notin Z$. We let L_Z be a line bundle with a meromorphic section s with divisor

$$(s) = Z.$$

Certain relations will involve Z linearly, and consequently to prove such relations we may assume that Z is effective, and so s is holomorphic. We suppose ρ is a hermitian metric on L_Z.

In practice, we shall be given a holomorphic map $f \colon \mathbf{C}^n \to X$ into a complex manifold, and the divisor Z will be the pull back of a divisor on X. Since X is assumed a manifold, this implies that Z is defined as a Cartier divisor, and the metric on L_Z is defined as the pull back of a metric on a line bundle on X.

So suppose Z is effective. One way to define a tubular neighborhood of $Z \cap \mathbf{B}(r)$ is to take the set

$$V(Z,\varepsilon)(r) = \{z \in \mathbf{B}(r) \text{ such that } |s(z)|_\rho^2 < \varepsilon\}.$$

Locally in the neighborhood of a point, s can be represented by a holomorphic function g, and $|s|_\rho^2 = |g|^2 h$ where h is positive and C^∞. Then locally, $V(Z, \varepsilon)$ consists of those $z \in \mathbf{B}(r)$ such that $|g(z)| < \varepsilon/h(z)$. For all but a discrete set of sufficiently small ε, this neighborhood has a regular boundry. We let its boundry be

$$S(Z, \varepsilon)(r) = \{z \in \mathbf{B}(r) \text{ such that } |s(z)|_\rho^2 = \varepsilon\}.$$

Letting r_1 be a number slightly bigger than r, the boundry of $\mathbf{B}(r) - V(Z, \varepsilon)(r)$ consists of

$$\mathbf{S}(r) - V(Z, \varepsilon)(r_1) \qquad \text{and} \qquad S(Z, \varepsilon)(r) \quad \text{(suitably oriented)}.$$

Note that the tubular neighborhood could also be defined using a reduced section s_0 defining the reduced divisor Z_0 rather than Z itself. In the following application of Stoke's theorem, all we are using is a family of tubular neighborhoods $S(Z, \varepsilon)$, depending on a parameter ε, such that this family shrinks to Z as ε approaches 0.

Theorem 3.1. *Let α be a C^∞ function on $\mathbf{C}^n - Z$, and let β be a C^∞ form, of type $(n-1, n-1)$ except on a negligible set of points on the boundary of $\mathbf{S}(r) - Z$. Assuming that all subsequent integrals are absolutely convergent, we have*

$$\int_{\mathbf{B}(r)} (dd^c\alpha \wedge \beta - \alpha \wedge dd^c\beta) = \int_{\mathbf{S}(r)} (d^c\alpha \wedge \beta - \alpha \wedge d^c\beta)$$

$$- \lim_{\varepsilon \to 0} \int_{S(Z,\varepsilon)(r)} (d^c\alpha \wedge \beta - \alpha \wedge d^c\beta).$$

Proof: We have

$$d(d^c\alpha \wedge \beta) = dd^c\alpha \wedge \beta - d^c\alpha \wedge d\beta$$
$$d(\alpha \wedge d^c\beta) = d\alpha \wedge d^c\beta + \alpha \wedge dd^c\beta.$$

Since β is of type $(n-1, n-1)$, we get $\partial\alpha \wedge \partial\beta = 0$ and $\bar{\partial}\alpha \wedge \bar{\partial}\beta = 0$, so

$$d^c\alpha \wedge d\beta = d^c\beta \wedge d\alpha$$

Applying Stokes and subtracting the two expressions above yields the theorem.

Note that we did not assume β closed, so that in the proof we can apply the technique of multiplying β by a function with appropriate small support in the neighborhood of a negligible singularity. However, in applications, we shall deal only with the case when β is closed, and more specially when $\beta = \omega^{n-1}$. Indeed, since $d\omega = 0$, we also get

$$d\omega^{n-1} = 0.$$

We thus specialize Stokes' theorem to the special case which we shall use:

If α is an admissible function, then for all but a discrete set of r:

$$\int_{B(r)} dd^c\alpha \wedge \omega^{n-1} = \int_{S(r)} d^c\alpha \wedge \omega^{n-1} - \lim_{\varepsilon \to 0} \int_{S(Z,\varepsilon)(r)} d^c\alpha \wedge \omega^{n-1}$$

and similarly with φ^{n-1} instead of ω^{n-1}.

In order to evaluate the limit on the right, we shall need a specific computation which determines the limit in each case. Here again, although we apply the computation to the case when $\beta = \omega^{n-1}$, we must carry out the statement with a general β, since we wish to multiply β by suitable functions in the proof.

The limit on the right will be called the **singular term** in Stokes' theorem.

We now compute the limit on the right for each case of admissible functions.

Lemma 3.2. *If α is C^∞ then*

$$\lim_{\varepsilon \to 0} \int_{S(Z,\varepsilon)(r)} d^c\alpha \wedge \beta = 0.$$

Proof: In this case, by Stoke's theorem,

$$\int_{S(Z,\varepsilon)(r)} d^c\alpha \wedge \beta = \int_{V(Z,\varepsilon)(r)} dd^c\alpha \wedge \beta - d^c\alpha \wedge d\beta,$$

and since the volume of $V(Z,\varepsilon)$ shrinks to 0 as $\varepsilon \to 0$, it follows that the right hand side approaces 0, whence the lemma follows.

The next result is classical, and is the standard case involving singularities which are not negligible. First we make a preliminary remark. Over a ball $\mathbf{B}(r_1)$ for r_1 slightly bigger than r, we can write the divisor as a Weil divisor

$$Z = \sum m_i Z_i$$

where the components Z_i are reduced and irreducible, and m_i are positive integers. Then an integral over Z is defined as a sum of integrals

$$\int_Z = \sum m_i \int_{Z_i},$$

and each integral over Z_i is defined as an integral over the regular points of Z_i. Note that the intersections of two or more components has complex codimension at least 2, so real codimension at least 4, whence this intersection is negligible from the point of view of Stoke's theorem.

Lemma 3.3 (Poincaré). *Let β be as in the theorem, and s holomorphic, defining the divisor Z. Let $S(Z,\varepsilon)$ be defined by a reduced section s_0 of L_Z as mentioned above. Then*

$$\lim_{\varepsilon \to 0} \int_{S(Z,\varepsilon)(r)} d^c \log |s|^2_\rho \wedge \beta = \int_{Z(r)} \beta.$$

74

Proof: Using a partition of unity, and the absolute convergence of the integrals which is assumed, we reduce the proof of the formula to the case when β has support at a regular point of Z, so that in a neighborhood of that point, s can be represented by a holomorphic function g, and s_0 can be represented by a complex coordinate function w_1, the first of a system of complex coordinates $w_1, \ldots, w_n$. The neighborhood can be chosen small enough that we can write

$$g(w) = w_1^m g_1(w)$$

where g_1 does not vanish at $w_1 = 0$ (defining Z locally). Then

$$|s|_\rho^2 = |w_1|^{2m} h(w)$$

where h is positive C^∞. In this case, putting $w_1 = ue^{i\theta}$ we have on the circle $u = \varepsilon$:

$$d^c \log |w_1|^2 = u \frac{\partial}{\partial u} \log u \frac{d\theta}{2\pi} = \frac{d\theta}{2\pi},$$

and $\log h$ is locally C^∞, so we can apply Lemma 3.2 to the term coming from the $d^c \log h$ contribution. As to the other term, we get

$$\lim_{\varepsilon \to 0} \int_{S(Z,\varepsilon)} d^c \alpha \wedge \beta = \lim_{\varepsilon \to 0} \int_{|w_1|=\varepsilon} d^c \log |w_1|^{2m} \wedge \beta$$

$$= m \int_{w_1=0} \beta = \int_{Z(r)} \beta,$$

which proves Poincaré's lemma.

The next case occurs when singularities are weak enough so that the singular term in Stokes' theorem vanishes.

Lemma 3.4. *Let α be of type III. Then the singular term in Stokes' formula vanishes; that is for any C^∞ form β of type $(n-1, n-1)$*

we have

$$\lim_{\varepsilon \to 0} \int_{S(Z,\varepsilon)(r)} d^c\alpha \wedge \beta = 0.$$

Proof: Again by using a partition of unity on β and the absolute convergence of the integrals involved, we are reduced to a local evaluation, when β has support in a small neighborhood of a point of Z, and α has a representation

$$\alpha = \log(1 + u^\lambda) \quad \text{where} \quad u = |g|^2 h,$$

where g is holomorphic, and h is positive C^∞. One could again argue as before, but I shall us an argument which Stoll uses [St], namely for $\delta > 0$, the function

$$\alpha_\delta = \log\left(1 + (\delta + u)^\lambda\right)$$

is C^∞ and we can apply Lemma 3.2. But

$$d^c\alpha_\delta = \frac{\lambda(\delta + u)^{\lambda-1} d^c u}{1 + (\delta + u)^\lambda}.$$

One can then take the limit under the integral sign as $\delta \to 0$ to conclude the proof.

Remark: Instead of β in the lemma, we may use ω^{n-1}. For one thing, the arguments still work for a form with negligible singularity at the origin. But also we assumed that $0 \notin Z$, so multiplying ω^{n-1} with a C^∞ function which is equal to 1 outside a small neighborhood of the origin does not change the validity of the formula to be proved, and reduces it to the case of the C^∞ form β as stated.

If we combine Lemma 3.4 with Stokes' theorem and Theorem 2.3, then we have proved:

Theorem 3.5. *Suppose $\alpha = \log \gamma$ is of type I or III, and is C^∞ near the origin. Then*

$$\int_0^r \frac{dt}{t} \int_{B(t)} dd^c \log \gamma \wedge \omega^{n-1} = \frac{1}{2} \int_{S(r)} (\log \gamma)\sigma - \frac{1}{2} \log \gamma(0).$$

For functions of mixed type I and II, we do get a singular contribution as follows.

Theorem 3.6. *Let Z be a divisor on $\mathbf{C}^n$ and let s be a meromorphic section of L_Z such that $(s) = Z$. Suppose $0 \notin Z$. Let ρ be a hermitian metric on L_Z and let $\gamma = |s|_\rho^2$. Then*

$$\int_0^r \frac{dt}{t} \int_{B(t)} dd^c \log \gamma \wedge \omega^{n-1} + \int_0^r \frac{dt}{t} \int_{Z(t)} \omega^{n-1}$$

$$= \frac{1}{2} \int_{S(r)} (\log \gamma)\sigma - \frac{1}{2} \log \gamma(0).$$

Proof: This is immediate from Lemmas 3.2 and 3.3, also using Theorem 2.3.

We shall use Theorems 3.5 and 3.6 in products and inverses of functions as in those theorems, and we use the homomorphic property of

$$\gamma \mapsto dd^c \log \gamma$$

to get relations in the next section.

II, §4. THE NEVALINNA FUNCTIONS AND THE FIRST MAIN THEOREM

In this section we introduce the definitions of the three basic Nevanlinna functions, and translate Theorems 3.5 and 3.6 into the new terminology. We are here in the equidimensional case, which stemmed from Carlson-Griffiths [C-G]. We shall make additional historical remarks later. We start by making the definitions on $\mathbf{C}^n$, and then get the theorems for any non-degenerate holomorphic map

$$f: \mathbf{C}^n \to X$$

into a compact complex manifold of dimension n by pull back.

The Nevanlinna functions on $\mathbf{C}^n$

Let Z be a divisor on $\mathbf{C}^n$, such that $Z = (g)$ where g is a meromorphic function, $0 \notin Z$. We define the **pre counting function** and **counting function** by

$$\mathbf{n}_Z(t) = \int_{Z(t)} \omega^{n-1} \quad \text{and} \quad N_Z(r) = \int_0^r \mathbf{n}_Z(t) \frac{dt}{t}$$

For a positive function γ sufficiently smooth such that the integrals are absolutely convergent, we define the **pre proximity function** and **proximity function** by

$$\mathbf{m}_\gamma^0(t) = \int_{S(t)} d^c \log \gamma \wedge \omega^{n-1} \quad \text{and} \quad m_\gamma^0(r) = \int_0^r \mathbf{m}_\gamma^0(r) \frac{dt}{t}$$

If $\log \gamma$ is admissible, then by Theorem 2.3,

$$m_\gamma^0(r) = \frac{1}{2} \int_{S(r)} (\log \gamma) \sigma - \frac{1}{2} \log \gamma(0).$$

Suppose that we define a metric ρ on the line bundle L_Z having the meromorphic section s represented by g, by the formula

$$|s|_\rho^2 = |g|^2 h$$

where h is a positive C^∞ function. We then take $\gamma = |s|_\rho^{-2}$ and we use the notation

$$m_{Z,\rho}^0(r) = \int_0^r \frac{dt}{t} \int_{S(t)} - d^c \log |s|_\rho^2 \wedge \omega^{n-1}.$$

Thirdly, we define the **pre height** and **height functions** associated to a $(1,1)$ form η by

$$t_\eta(t) = \int_{B(t)} \eta \wedge \omega^{n-1} \quad \text{and} \quad T_\eta(r) = \int_0^r \frac{dt}{t} \int_{B(t)} \eta \wedge \omega^{n-1}.$$

This definition will be applied to the special case when $\eta = dd^c \log \gamma$ for suitable functions γ, in which case we also use the notation

$$\text{Ric } \gamma = dd^c \log \gamma \quad \text{and so} \quad T_{\text{Ric } \gamma}(r) = \int_0^r \frac{dt}{t} \int_{B(t)} dd^c \log \gamma \wedge \omega^{n-1}.$$

Thus we use the abbreviation $\text{Ric } \gamma$ instead of $\text{Ric}(\gamma\Phi)$, according to our definition of the Ricci form in §1. We may also call T_η the **height transform of** η. For the height, recall that we defined the Chern form of a metric ρ by

$$c_1(\rho) = -dd^c \log h.$$

Then we also use the notation

$$T_\rho(r) = \int_0^r \frac{dt}{t} \int_{B(t)} c_1(\rho) \wedge \omega^{n-1}.$$

79

Observe that each one of the proximity and height functions are homomorphic in γ, in other words

$$\gamma \mapsto m_\gamma^0 \quad \text{and} \quad \gamma \mapsto T_{\text{Ric } \gamma}$$

sends products into sums. The counting function is homomorphic in Z, that is

$$Z \mapsto N_Z$$

is a homomorphism. For the height,

$$\rho \mapsto T_\rho$$

is a homomorphism, where on the left we take the tensor product of metrics on the tensor product of line bundles.

We may now reformulate Theorem 3.6 in terms of the terminology we have just defined.

Theorem 4.1 *Let Z be a divisor on $\mathbf{C}^n$ and let s be a meromorphic section s of L_Z. Let ρ be a metric on L_Z. Suppose $0 \notin Z$. Then*

$$t_\rho = n_Z + m_{Z,\rho}^0, \quad \text{and} \quad T_\rho = N_Z + m_{Z,\rho}^0.$$

The Nevanlinna functions for a mapping $f: \mathbf{C}^n \to X$

Consider now a holomorphic map

$$f: \mathbf{C}^n \to X$$

which is non-degenerate, i.e., is a local isomorphism somewhere, so its image is not contained in any divisor of the compact complex manifold X. If D is a divisor on X, s a meromorphic section of L_D with $(s) = D$, and ρ is a metric on L_D then we can define the three Nevanlinna functions by pull back.

80

Pre height and height

$$t_{f,\rho}(r) = \int_{B(r)} f^* c_1(\rho) \wedge \omega^{n-1} \quad \text{and} \quad T_{f,\rho}(r) = \int_0^r t_{f,\rho}(t)\frac{dt}{t}$$

If η is a $(1,1)$ form on X, we also use the same notation

$$t_{f,\eta}(r) = \int_{B(r)} f^* \eta \wedge \omega^{n-1} \quad \text{and} \quad T_{f,\eta} = \int_0^r t_{f,\eta}(t)\frac{dt}{t}.$$

Pre proximity and proximity functions

$$m^0_{f,D,\rho}(r) = \int_{S(r)} - d^c \log |s \circ f|^2_\rho \wedge \omega^{n-1} \quad \text{and}$$

$$m^0_{f,D,\rho}(r) = \int_0^r m^0_{f,D,\rho}(t)\frac{dt}{t}.$$

Then

$$m^0_{f,D,\rho}(r) = \frac{1}{2} \int_{S(r)} (\log |s \circ f|^{-2}_\rho)\sigma + \frac{1}{2}\log |s \circ f(0)|^2_\rho.$$

Counting functions

$$n_{f,D}(r) = \int_{f^*D(r)} \omega^{n-1} \quad \text{and} \quad N_{f,D}(r) = \int_0^r n_{f,D}(t)\frac{dt}{t}.$$

Proposition 4.2. *Let ρ, ρ' be two metrics on the line bundle L on X. Then*

$$T_{f,\rho'} = T_{f,\rho} + O(1).$$

In other words, two height functions associated with a line bundle differ by a bounded function.

Proof: There is a positive C^∞ function $\gamma > 0$ such that $\rho' = \gamma\rho$. Then $c_1(\rho)$ and $c_1(\rho')$ differ by the form $dd^c \log \gamma$, and $\log \gamma$ is bounded because it is a continuous function on the compact space X. Pulling back to $\mathbf{C}^n$, and using Stokes' theorem in its simplest form when the singular divisor is 0, we get

$$T_{f,\rho} - T_{f,\rho'} = \int_0^r \frac{dt}{t} \int_{B(t)} dd^c \log \gamma \wedge \omega^{n-1} = \frac{1}{2} \int_{S(r)} (\log \gamma)\sigma - \frac{1}{2} \log \gamma(0).$$

Since $\log \gamma$ is bounded, the right hand side is bounded, thus proving the proposition.

Proposition 4.3. *Let η, η' be $(1,1)$ forms, and assume η positive. Then*

$$T_{f,\eta'} = O(T_{f,\eta}).$$

In particular, if η, η' are both positive, then

$$T_{f,\eta'} \gg\ll T_{f,\eta}.$$

Proof: Since X is compact, there exist a constant k such that

$$\eta' \le k\eta$$

so the proposition is obvious.

Remark: The proposition means that the heights associated with positive $(1,1)$ forms are of the same order of magnitude. In some applications we shall take their logs, and so we get

$$\log T_{f,\eta'} = \log T_{f,\eta} + O(1).$$

In particular, if we are interested only in the functions mod $O(1)$ (multiplicatively or additively), we may simply write T_f or $\log T_f$ respectively, without any further subscript. In the one-dimensional theory, one usually writes

$$T_f = T_{f,\infty}$$

if f is a holomorphic map into the projective line, but one is only interested in the order of magnitude of T_f in certain estimates.

Theorem 4.4. *For any metric ρ on L_D we have*

$$T_{f,\rho} = N_{f,D} + m^0_{f,D,\rho}.$$

Proof: This is simply a restatement of Theorem 4.1, taking the pull back into account. Of course, the theorem also holds at the pre level, that

$$\mathbf{t}_{f,\rho} = \mathbf{n}_{f,D} + \mathbf{m}^0_{f,D,\rho}$$

as functions defined for all but a discrete set of values r.

Let Ω be a volume form on X. Let us write

$$f^*\Omega = \gamma_f \Phi, \quad \text{with} \quad \gamma_f = \gamma_{f,\Omega},$$

where Φ is the euclidean volume form on $\mathbf{C}^n$ as defined in §1. Then locally,

$$\gamma_f = |\Delta|^2 h$$

where Δ is holomorphic, $h > 0$ is C^∞, and locally, (Δ) is the **ramification divisor**

$$(\Delta) = \mathrm{Ram}_f$$

defined locally by the Jacobian determinant of f in terms of complex coordinates. The volume form may be viewed as defining a metric κ on the canonical bundle, and we also define the **Ricci form**

$$\mathrm{Ric}\,\Omega = c_1(\kappa).$$

The pull back of the Chern form of this metric is given outside Ram_f by

$$f^* \mathrm{Ric}\, \Omega = f^* c_1(\kappa) = dd^c \log \gamma_f = dd^c \log h.$$

Let us define the **height associated with the volume form** Ω to be

$$T_{f,\mathrm{Ric}\,\Omega} = T_{f,\kappa} = \int_0^r \frac{dt}{t} \int_{\mathbf{B}(t)} dd^c \log \gamma_f \wedge \omega^{n-1}.$$

As a special case of Theorem 4.1, actually Theorem 3.6 plus notation, we then get:

Theorem 4.5.

$$T_{f,\kappa}(r) + N_{f,\mathrm{Ram}}(r) = \frac{1}{2} \int_{\mathbf{S}(r)} (\log \gamma_f)\sigma - \frac{1}{2} \log \gamma_f(0).$$

If D denotes a divisor on X, we denote by $T_{f,D}$ the height function $T_{f,\rho}$ for any metric ρ on L_D. By Proposition 4.2, $T_{f,D}$ is well defined mod $O(1)$, i.e. mod bounded functions. Thus we could write $T_{f,K}$ instead of $T_{f,\kappa}$ if K is a canonical divisor, with the understanding that the relation of Theorem 4.4 then holds with an added term $O(1)$. In most applications, one is looking only for relations mod $O(1)$.

So far in this section, we have had formal relationships derived from an application of Stokes' theorem and a computation of the singular term. In the next section, we estimate terms like those occurring on the right hand side in Theorem 4.4, and activate another technique.

II, §5. THE CALCULUS LEMMA

The calculus lemma will be applied in the same way as before, but with dt/t replaced by dt/t^{2n-1}. We use Lemma 3.1 of Chapter I with the function ψ unchanged, however.

Lemma 5.1. *Let F be a function of one variable $r > 0$ such that the first derivative exists, and F' is piecewise of class C^1. Suppose that both $F(r)$ and $r^{2n-1}F'(r)$ are positive increasing functions of r, and that there exists r_1 such that $F(r_1) \geq e$ for $r \geq r_1$. Let $b_1 \geq 1$ be the smallest number such that*

$$b_1 r^{2n-1} F'(r) \geq e \quad \text{for all} \quad r \geq 1.$$

Then

$$\frac{1}{r^{2n-1}} \frac{d}{dr} \left(r^{2n-1} \frac{dF}{dr} \right) \leq F(r)\psi(F(r))\psi(r^{2n-1}b_1 F(r)\psi(F(r)))$$

for all $r \geq r_1$ outside a set of measure $\leq 2b_0(\psi)$.

The proof is again by a double application of Lemma 3.1 of Chapter I. For our purposes now, we define the **error term**

$$S(F, b_1, \psi, r) = \log\{F(r)\psi(F(r))\psi(r^{2n-1}b_1 F(r)\psi(F(r)))\},$$

i.e. the log of the right hand side in the above inequality for the repeated derivative.

We shall obtain a function F to which we apply the inequality as follows.

Suppose α is a function which is positive and continuous except on a divisor of $\mathbf{C}^n$. We shall need that α is continuous at 0 for convenience as usual, but also that

$$t \mapsto \int_{S(t)} \alpha\sigma$$

85

is piecewise continuous. I leave to a future write up what I hope will be a neat characterization of such α, but I don't know how to do it now.

For such α, we define the **height transform**

$$F_\alpha(r) = \int_0^r \frac{dt}{t^{2n-1}} \int_{B(t)} \alpha\Phi, \quad \text{where} \quad \Phi = \prod \frac{\sqrt{-1}}{2\pi} dz_i \wedge d\bar{z}_i.$$

Then first F_α is differentiable, with derivative

$$F_\alpha'(r) = \frac{1}{r^{2n-1}} \int_{B(r)} \alpha\Phi;$$

and using spherical coordinates in higher dimension, writing $\alpha(t, u)$ where u is the variable on the sphere of radius 1, we have

$$r^{2n-1} F_\alpha'(r) = \frac{2}{(n-1)!} \int_0^r dt \int_{S(1)} \alpha(t, u) t^{2n-1} \sigma(u).$$

Hence

$$\frac{d}{dr}\left(r^{2n-1}\frac{dF_\alpha}{dr}\right) = \frac{2}{(n-1)!} r^{2n-1} \int_{S(r)} \alpha\sigma.$$

Hence finally,

$$\frac{1}{r^{2n-1}} \frac{d}{dr}(r^{2n-1} F_\alpha'(r)) = \frac{2}{(n-1)!} \int_{S(r)} \alpha\sigma.$$

Thus F_α satisfies the hypotheses of Lemma 5.1, and by that lemma, we find:

Proposition 5.2. *We have the inequality*

$$\log \int_{S(r)} \alpha\sigma \leq S(F_\alpha, b_1, \psi, r) + \log \frac{(n-1)!}{2}$$

for all $r \geq r_1(F_\alpha)$ outside a set of measure $\leq 2b_0(\psi)$.

II, §6. THE TRACE AND DETERMINANT IN THE MAIN THEOREM

So far, the theory in higher dimension has essentially been completely parallel to the one dimensional case. We now meet an additional phenomenon, where the higher dimension plays a role which was not noticeable before.

Let η be a $(1,1)$ form on an open set of $\mathbf{C}^n$. We can write

$$\eta = \sum \eta_{ij} \frac{\sqrt{-1}}{2\pi} dz_i \wedge d\bar{z}_j. \cdot$$

We define the **trace** and **determinant** of η by

$$\mathrm{tr}(\eta) = \sum \eta_{ii} \quad \text{and} \quad \det \eta = \det(\eta_{ij}).$$

Trivially from the definitions, we get the coordinate free expression

$$\boxed{(\det \eta)\Phi = \frac{1}{n!}\eta^n.}$$

To study the trace further, we introduce a new form, namely

$$\varphi(z) = dd^c \|z\|^2 = \sum \frac{\sqrt{-1}}{2\pi} dz_i \wedge d\bar{z}_i.$$

Then

$$\varphi^{n-1} = (n-1)! \sum \left(\frac{\sqrt{-1}}{2\pi}\right)^{n-1} dz_1 \wedge d\bar{z}_1 \wedge \cdots \wedge \widehat{dz_j \wedge d\bar{z}_j} \wedge \cdots \wedge dz_n \wedge d\bar{z}_n.$$

Therefore we get a neat expression for the trace:

$$\boxed{\eta \wedge \varphi^{n-1} = (n-1)!\,\mathrm{tr}(\eta)\Phi.}$$

Next, we show how the height can be defined in terms of φ instead of the form ω which we used previously.

First remark that *restricted to the sphere* $S(r)$,

$$d^c \|z\|^2 = r^2 d^c \log \|z\|^2 \quad \text{and} \quad dd^c \|z\|^2 = r^2 dd^c \log \|z\|^2.$$

Both relations are immediate.

Proposition 6.1. *If η is a closed C^∞ $(1,1)$ form, then*

$$r^{2n-2} \int_{B(r)} \eta \wedge \omega^{n-1} = \int_{B(r)} \eta \wedge \varphi^{n-1}.$$

The same relation holds if $\eta = dd^c \alpha$, where α is C^∞ except for singularities on a divisor such that the singular term in Stokes' theorem vanishes, as in Lemma 3.4.

Proof: If $n = 1$ the relation is trivial. Let $n \geq 2$. In the C^∞ case, we have:

$$r^{2n-2} \int_{B(r)} \eta \wedge \omega^{n-1} = r^{2n-2} \int_{B(r)} d(\eta \wedge d^c \log \|z\|^2 \wedge \omega^{n-2})$$

$$= r^{2n-2} \int_{S(r)} \eta \wedge d^c \log \|z\|^2 \wedge \omega^{n-2}$$

$$= \int_{S(r)} \eta \wedge d^c \|z\|^2 \wedge \varphi^{n-2}$$

$$= \int_{B(r)} \eta \wedge \varphi^{n-1}.$$

This proves the proposition when η is C^∞. Now suppose $\eta = dd^c \alpha$, where α is such that the singular term vanishes in Stokes' theorem.

Then we have directly

$$r^{2n-2} \int_{\mathbf{B}(r)} dd^c\alpha \wedge \omega^{n-1} = r^{2n-2} \int_{\mathbf{S}(r)} d^c\alpha \wedge \omega^{n-1}$$

$$= \int_{\mathbf{S}(r)} d^c\alpha \wedge \varphi^{n-1}$$

$$= \int_{\mathbf{B}(r)} \eta \wedge \varphi^{n-1}$$

thus concluding the proof.

Our first application is to the C^∞ case, to give another formula for the height in certain cases.

Let as before X be a compact complex manifold, and let η be a closed positive $(1,1)$-form. We let

$$\Omega = \frac{1}{n!}\eta^n$$

so Ω is a volume form. We let $f: \mathbf{C}^n \to X$ be a non-degenerate map, and

$$f^*\Omega = \gamma_f \Phi \qquad \text{so} \qquad \gamma_f = \det(f^*\eta).$$

The next proposition shows how the height is given by a height transform.

Proposition 6.2. *Let $\tau_f = \text{tr } f^*\eta$. Then*

$$T_{f,\eta} = (n-1)! F_{\tau_f}.$$

Proof: From the definitions and Proposition 6.1, we find:

$$T_{f,\eta}(r) = \int_0^r \frac{dt}{t} \int_{\mathbf{B}(t)} f^*\eta \wedge \omega^{n-1}$$

$$= \int_0^r \frac{dt}{t^{2n-1}} \int_{\mathbf{B}(t)} f^*\eta \wedge \varphi^{n-1}$$

$$= (n-1)! \int_0^r \frac{dt}{t^{2n-1}} \int_{\mathbf{B}(t)} (\operatorname{tr} f^*\eta)\Phi = (n-1)!F_{\tau_f}(r)$$

as was to be shown.

We can now pick up the relation of Theorem 4.4 and combine it with the calculus lemma and the expression of the height as a height transform, to get the Second Main Theorem when the divisor D is 0, so as in the one dimensional case, this amounts to a theorem about the ramification divisor.

Remark: *If η is hermitian semipositive, then*

$$(\det \eta)^{1/n} \leq \frac{1}{n}\operatorname{tr} \eta.$$

This is immediate, since the relation is one of linear algebra at each point, and so the relation follows immediately after diagonalizing the matrix of the form.

Theorem 6.3. *Let η be a closed, positive $(1,1)$-form, with $\Omega = \eta^n/n!$, and let $T_{f,\kappa}$ be the height associated with the volume form Ω, so $T_{f,\kappa} = T_{f,\operatorname{Ric}\Omega}$. Then*

$$T_{f,\kappa}(r) + N_{f,\mathrm{Ram}}(r) \leq \frac{n}{2}S(T_{f,\eta}, b_1, \psi, r) - \frac{1}{2}\log \gamma_f(0)$$

for all $r \geq r_1$ outside a set of measure $\leq 2b_0(\psi)$, where

$$r_1 = r_1\left(\frac{1}{(n-1)!}T_{f,\eta}\right) \qquad and \qquad b_1 = b_1\left(\frac{1}{(n-1)!}T_{f,\eta}\right).$$

Proof: By the remark,

$$\gamma_f^{1/n} = (\det f^*\eta)^{1/n} \leq \frac{1}{n}\operatorname{tr} f^*\eta.$$

We then estimate the right hand side in Theorem 4.5. We get:

$$\frac{1}{2}\int_{\mathbf{S}(r)} (\log \gamma_f)\sigma = \frac{n}{2}\int_{\mathbf{S}(r)} (\log \gamma_f^{1/n})\sigma \leq \frac{n}{2}\log \int_{\mathbf{S}(r)} \gamma_f^{1/n}\sigma$$

$$\leq \frac{n}{2}\log \int_{\mathbf{S}(r)} \tau_f\sigma$$

$$\text{[by Proposition 5.2]} \quad \leq \frac{n}{2}S(F_{\tau_f}, b_1, \psi, r) + \log(n-1)!$$

$$\text{[by Proposition 6.2]} \quad \leq \frac{n}{2}S(T_{f,\eta}, b_1, \psi, r)$$

for $r \geq r_1$ outside the exceptional set, thus proving the theorem.

Remark: By Proposition 4.3, we know that two height functions associated with positive $(1,1)$ forms are of the same order of magnitude. Since the error term $S(T_{f,\eta}, \psi, r)$ grows slowly, essentially like a log, we may replace $T_{f,\eta}$ by any constant multiple of $T_{f,\eta}$, at the cost of adding $O(1)$ at the end of the estimate.

The theorem with the usual error term $O(\log r + \log T_{f,\eta}(r))$ is due to Carlson-Griffiths. The improvement of the error term with the precise factor $n/2$ and the arbitrary estimating function ψ is due to Lang [**La 8**], see [**Wo**] who followed my suggestion. As before, it is not known if the factor $n/2$ is best possible in general.

II, §7. A GENERAL SECOND MAIN THEOREM (AHLFORS-WONG METHOD)

As in the one dimensional case, we formulate and prove a Second Main Theorem. The method stems partly originally from F. Nevanlinna, who was the first to use a singular volume form in the one-dimensional case. Then Carlson-Griffiths carried out the higher dimensional case [C-G], while P.M. Wong [Wo] used a different singular volume form making more precise a method of Ahlfors. We shall make additional historical comments at the end.

Let X as before be a compact complex manifold. Let D be a divisor on X. We say that D has **simple normal crossings** if

$$D = \sum_j D_j$$

where each D_j is irreducible, non-singular, and at each point of X there exist complex coordinates $z_1, \ldots, z_n$ such that D in a neighborhood of this point is defined by

$$z_1 \cdots z_k = 0 \qquad \text{with} \quad k \leq n.$$

When $n = 1$, then the property of D having simple normal crossings is equivalent to the property that D consists of distinct points, taken with multiplicity 1. The maximal value of k which can occur will be called the **complexity** of D. Throughout we consider a non-degenerate holomorphic map

$$f \colon \mathbf{C}^n \to X.$$

For the rest of this section we let:

$D = \sum D_j$ have simple normal crossings of complexity k;

$L_j = L_{D_j}$ be the line bundle associated with D_j, with a metric ρ_j;

$\Omega =$ volume form on X, defining a metric κ on the canonical bundle L_K so $c_1(\kappa) = \operatorname{Ric} \Omega$;

$\gamma_f =$ the function such that $f^*\Omega = \gamma_f \Phi$;

$\eta =$ a closed, positive $(1,1)$-form such that $c_1(\rho_j) \leq \eta$ for all j, and $\Omega \leq \eta^n/n!$.

92

Theorem 7.1. *Let f, D, ρ_j, κ, and η be as above. Suppose that $f(0) \notin D$ and $0 \notin \mathrm{Ram}_f$. Then*

$$T_{f,\kappa}(r) + \sum T_{f,\rho_j}(r) - N_{f,D}(r) + N_{f,\mathrm{Ram}}(r)$$
$$\leq \frac{n}{2} S(BT_{f,\eta}^{1+k/n}, b_1, \psi, r) - \frac{1}{2} \log \gamma_f(0) + 1$$

for all $r \geq r_1$ outside a set of measure $\leq 2b_0(\psi)$, and some constant $B = B(\Omega, D, \eta)$. The constants B, b_1, and r_1 will be determined explicitly below.

The general shape of the theorem stems from Carlson-Griffiths, but the good error term stems from Wong except for three improvements which I made on Wong's method: the use of the general function ψ; the exponent $1 + k/n$ instead of 2 (in this way, one recovers the essential estimate of [**La 8**]); and third, Wong takes the irreducible components of D in one linear system, but this is not necessary, and the statement of the theorem as well as its proof are easier without this extraneous assumption.

Note that the exponent $1 + k/n$ is also valid for $k = 0$, in which case we recover Theorem 6.3. This exponent therefore interpolates in the error term for the singularities of D. If D does not have normal crossings, it would be interesting to give an error term showing the dependence on the complexity of the singularities of D in this more general case.

We shall give Wong's proof, suitably adjusted. The proof is a refinement of Ahlfors' method [**Ah**] pp. 22-27, and also makes use of the curvature ideas of Carlson-Griffiths [**C-G**], but somewhat more efficiently.

As in the one-dimensional case, we observe that we do not necessarily need to work with a map of $\mathbf{C}^n$ into X. We could work with a map $f \colon \mathbf{B}(R) \to X$, and the final estimate gives an implicit bound on R in case the canonical class K is ample. In this case, there do not exist non-degenerate maps of $\mathbf{C}^n$ into X, but a bound on the radius R still provides a theorem.

For the rest of this section we let:

s_j = holomorphic section of L_j such that $(s_j) = D_j$; after multiplying
$\quad s_j$ by a small constant if necessary, we may assume without loss
$\quad$ of generality that

$$|s_j|_{\rho_j} \leq 1/e, \quad \text{say}$$

$|s_j \circ f|_j^2 = |s_j \circ f|_{\rho_j}^2$
Λ = positive decreasing function of r with $0 < \Lambda < 1$.

As in Wong [**Wo**] and Stoll [**St**], we define the **Ahlfors-Wong singular form** with constant λ to be

$$\Omega(D)_\lambda = \frac{\Omega}{\Pi|s_j|_j^{2(1-\lambda)}}.$$

If Λ is not constant, then we shun taking $dd^c \log$. But with $\Lambda = \lambda$ equal to a constant, then we shall take $dd^c \log$, and the only thing that happens is that the factor $(1 - \lambda)$ comes out in front. We let

$$\Omega(D)_{f,\Lambda} = \prod_j |s_j \circ f|_j^{-2(1-\Lambda)} f^*\Omega = \gamma_\Lambda \Phi$$

where
$$\gamma_\Lambda = \gamma_{f,D,\Lambda} = \frac{\gamma_f}{\Pi|s_j \circ f|_j^{2(1-\Lambda)}}$$

Then in particular, from the assumption $|s_j|_j \leq 1/e \leq 1$, we have

$$\gamma_f \leq \gamma_\Lambda.$$

Note that with non-constant Λ, $\Omega(D)_{f,\Lambda}$ is not the pullback of a form on X. As in dimension 1, we shall prove the fundamental estimate:

Proposition 7.2. *Let* $r_1 = r_1(F_{\gamma_f^{1/n}})$, *and let*

$$\Lambda(r) = \begin{cases} 1/qT_{f,\eta}(r) & \text{for } r \geq r_1 \\ \text{constant} & \text{for } r \leq r_1. \end{cases}$$

Let k *be the complexity of* D. *Then for some constants* B, b_1 *we have*

$$\frac{n}{2}\log \int_{S(r)} \gamma_\Lambda^{1/n}\sigma \leq \frac{n}{2}S(BT_{f,\eta}^{1+k/n}, b_1, \psi, r)$$

for $r \geq r_1$ *outside a set of measure* $\leq 2b_0(\psi)$.

Explicit determinations of the constants B and b_1 will be given below. Note that since η was chosen so that $\Omega \leq \eta^n/n!$, we get $F_{\gamma_f^{1/n}} \leq T_{f,\eta}/n!$. Therefore $r_1(F_{\gamma_f^{1/n}}) \geq r_1(T_{f,\eta}/n!)$. Hence we have $\Lambda < 1$. We postpone the proof of Proposition 7.2 and see immediately how the proposition implies the main theorem. First note that by assumption, for all j,

$$T_{f,\rho_j} \leq T_{f,\eta}$$

We apply Theorem 3.6 and the definitions, and we use the fact that $dd^c \log$ transforms multiplication to addition. *In the case when* λ *is constant,* $0 < \lambda < 1$, *we obtain:*

(1)
$$T_{f,\kappa}(r) + (1-\lambda)\sum T_{f,\rho_j}(r) - (1-\lambda)\sum N_{f,D_j}(r) + N_{f,\text{Ram}}(r)$$
$$= \frac{n}{2}\int_{S(r)} (\log \gamma_\lambda^{1/n})\sigma - \frac{1}{2}\log \gamma_\lambda(0).$$

We are interested in an inequality. Since $N_{f,D_j} \geq 0$, we can delete the factor $(1-\lambda)$ in front to get a smaller quantity on the left hand side. We now use the special value $\lambda = \Lambda(r) = 1/qT_{f,\eta}(r)$. Then the undesirable term on the left hand side satisfies

$$-1 \leq -\Lambda(r)\sum T_{f,\rho_j}(r)$$

95

by the assumption on η. For the constants, we have a bound independent of λ:

$$-\frac{1}{2} \log \gamma_\lambda(0) \le -\frac{1}{2} \log \gamma_f(0)$$

because $\gamma_f \le \gamma_\lambda$. So, the term on the right in Theorem 7.1 is

$$-\frac{1}{2} \log \gamma_f(0) + 1.$$

Finally we take the log outside the integral over $S(r)$ to get

$$\int_{S(r)} (\log \gamma_{\Lambda(r)}^{1/n}) \sigma \le \log \int_{S(r)} \gamma_\Lambda^{1/n} \sigma,$$

and we apply Proposition 7.2 to conclude the proof of Theorem 7.1.

There remains to give:

Proof of Proposition 7.2. We are set up to apply the calculus lemma to functions α and F_α with various choices for α, where as before

$$F_\alpha(r) = \int_0^r \frac{dt}{t^{2n-1}} \int_{B(r)} \alpha \Phi.$$

At first, we deal with an arbitrary Λ, not necessarily that of Proposition 7.2. Note that

$$F_{\gamma_\Lambda^{1/n}} \ge F_{\gamma_f^{1/n}} \qquad \text{and} \qquad F'_{\gamma_\Lambda^{1/n}} \ge F'_{\gamma_f^{1/n}}$$

So we take:

The constants b_1 and r_1: $b_1 = b_1(F_{\gamma_f^{1/n}})$ and $r_1 = r_1(F_{\gamma_f^{1/n}})$ are such that

$$F_{\gamma_\Lambda^{1/n}}(r) \ge e \qquad \text{and} \qquad b_1 r^{2n-1} F'_{\gamma_\Lambda^{1/n}}(r) \ge e \text{ for } r \ge r_1.$$

Thus we obtain the bound

$$(2) \qquad \frac{n}{2} \log \int_{S(r)} \gamma_\Lambda^{1/n} \sigma \le \frac{n}{2} S(F_{\gamma_\Lambda^{1/n}}, b_1, \psi, r) + \frac{n}{2} \log \frac{(n-1)!}{2}.$$

We shall prove below the following estimate (replacing a **curvature estimate**).

Lemma 7.3. *There is a constant* $b = b(\Omega, D, \eta)$ *such that for any decreasing* Λ *with* $0 < \Lambda < 1$ *we have*

$$F_{\gamma_\Lambda^{1/n}} \leq (q+1)\frac{b^{1/n}}{n!}\frac{1}{\Lambda^{k/n}}T_{f,\eta} + \frac{1}{2}\frac{b^{1/n}}{n!}q\log 2\frac{1}{\Lambda^{1+k/n}}.$$

In particular, if $\Lambda = 1/qT_{f,\eta}$ *as in Proposition 7.2, then for a suitable constant* B *we have*

$$F_{\gamma_\Lambda^{1/n}} \leq \frac{B}{(n-1)!}T_{f,\eta}^{1+k/n}.$$

Assume Lemma 7.3 for the moment, and take $\Lambda = 1/qT_{f,\eta}$ as in Proposition 7.2. Let B be whatever comes out of the lemma, namely

The constant B : $\qquad B = \dfrac{b^{1/n}}{n}((q+1)q^{k/n} + \frac{1}{2}q^{2+k/n}\log 2).$

Then indeed

$$F_{\gamma_\Lambda^{1/n}} \leq \frac{B}{(n-1)!}T_{f,\eta}^{1+k/n}.$$

Substituting this estimate in (2) concludes the proof of Theorem 7.2, with the specific constant B as above when $c_1(\rho_j) \leq \eta$ for all j.

There remains to be proved Lemma 7.3. We base the proof on still another lemma, which also gives us a value for the constant b.

Lemma 7.4. *There is a constant b, depending only on Ω, η and D, via the sections s_j and metrics ρ_j, such that if $0 < \lambda < 1$ and if we put*

$$\eta_{D,\lambda} = (q+1)\lambda\eta + \sum dd^c \log(1 + |s_j|_j^{2\lambda}),$$

then $\eta_{D,\lambda}$ is closed and > 0 outside D, and

$$\lambda^{n+k}\Omega(D)_\lambda \le b\frac{1}{n!}\eta_{D,\lambda}^n.$$

We shall first see how Lemma 7.4 implies Lemma 7.3. In the first place, it suffices to prove Lemma 7.3 with a fixed λ rather than Λ. Indeed, since Λ is a decreasing function, for a given value of r we have

$$\gamma_\Lambda(z) \le \gamma_{\Lambda(r)}(z) \quad \text{for} \quad |z| \le r,$$

so letting $\lambda = \Lambda(r)$ we get

$$F_{\gamma_\Lambda^{1/n}}(r) \le F_{\gamma_\lambda^{1/n}}(r),$$

which implies Lemma 7.3 with the possibly variable Λ.

Now we have by Lemma 7.4 taking the pull back f^*, the inequality

$$\lambda^{n+k}\gamma_\lambda\Phi \le b(\det f^*\eta_{D,\lambda})\Phi,$$

so that

$$\lambda^{1+k/n}\gamma_\lambda^{1/n}\Phi \le b^{1/n}(\det f^*\eta_{D,\lambda})^{1/n}\Phi$$

$$\le \frac{b^{1/n}}{n}(\operatorname{tr} f^*\eta_{D,\lambda})\Phi$$

$$\le \frac{b^{1/n}}{n!}f^*\eta_{D,\lambda} \wedge \varphi^{n-1}.$$

Therefore

$$\frac{n!}{b^{1/n}}\lambda^{1+k/n}F_{\gamma_\lambda^{1/n}}(r)$$

$$\leq \int_0^r \frac{dt}{t^{2n-1}} \int_{\mathbf{B}(t)} f^* \eta_{D,\lambda} \wedge \varphi^{n-1}$$

$$= \int_0^r \frac{dt}{t} \int_{\mathbf{B}(t)} f^* \eta_{D,\lambda} \wedge \omega^{n-1}$$

$$= (q+1)\lambda \int_0^r \frac{dt}{t} \int_{\mathbf{B}(t)} f^* \eta \wedge \omega^{n-1} + \sum \int_0^r \frac{dt}{t} \int_{\mathbf{B}(t)} dd^c \log(1+|s_j \circ f|_j^{2\lambda}) \wedge \omega^{n-1}$$

$$= (q+1)\lambda T_{f,\eta}(r) + \sum \frac{1}{2} \int_{\mathbf{S}(r)} \log(1+|s_j \circ f|_j^{2\lambda})\sigma - \sum \frac{1}{2}\log(1+|s_j \circ f(0)|_j^{2\lambda})$$

(by Stokes' theorem and Lemma 3.5 in Case III)

$$\leq (q+1)\lambda T_{f,\eta}(r) + \frac{1}{2}q\log 2$$

(because the expression inside the log is ≤ 2). This proves Lemma 7.3 with λ instead of Λ, but as we have already remarked, it also proves Lemma 7.3 in full.

Proof of Lemma 7.4. This is the curvature computation. We recall the general formula already given in Chapter I, §5 in a similar context:

$$(3) \qquad dd^c \log(1+u) = \frac{u\,dd^c \log u}{1+u} + \frac{1}{u(1+u)^2}du \wedge d^c u.$$

Here we let $u = u_j$ be the function

$$0 \leq u_j = |s_j|_j^{2\lambda} < 1 \qquad \text{so that} \qquad \frac{u_j\,dd^c \log u_j}{(1+u_j)} = -\frac{\lambda u_j}{1+u_j}c_1(\rho_j).$$

99

Since $\eta \geq c_1(\rho_j)$ we also have $\lambda\eta \geq \lambda u_j(1+u_j)^{-1}c_1(\rho_j)$, and so

$$\eta_{D,\lambda} \geq \lambda\eta + \sum\{\frac{\lambda u_j}{1+u_j}c_1(\rho_j) + dd^c\log(1+u_j)\}$$

$$\text{(4)} \qquad = \lambda\eta + \sum \frac{1}{u_j(1+u_j)^2}du_j \wedge d^c u_j \qquad \text{by (3)}$$

In particular, the right hand side is ≥ 0, so we have proved the positivity of $\eta_{D,\lambda}$. We shall now use systematically that if $\eta_1, \ldots, \eta_n$ are positive $(1,1)$ forms, then $\eta_1 \wedge \cdots \wedge \eta_n$ is also positive. Then given a point $x \in X$ which does not lie on the divisor D, we can find a small enough neighborhood U of x and a constant b_U such that

$$\eta_{D,\lambda}^n|U \geq b_U\lambda^n\Omega(D)_\lambda|U$$

because η is positive and $du_j \wedge d^c u_j \geq 0$, so we may use merely the term $(\lambda\eta)^n$ in the expansion of $\eta_{D,\lambda}^n$.

Suppose next that x lies on the divisor. We pick a small neighborhood U with complex coordinates $z_1, \ldots, z_n$ such that D is defined by

$$z_1 \cdots z_k = 0 \quad \text{with} \quad k \leq n.$$

Let

$$\eta_j = \frac{1}{u_j}du_j \wedge d^c u_j \geq 0.$$

Then by (4) and using the fact $(1+u_j)^2 \leq 4$, shrinking U further if necessary, there exists a constant $b_4 = b_4(\eta, U)$ such that

$$\eta_{D,\lambda}|U \geq \lambda b_4 \sum_{i=1}^{n} \frac{\sqrt{-1}}{2\pi}dz_i \wedge d\bar{z}_i + \frac{1}{4}\sum_{j=1}^{k} \eta_j$$

$$= \lambda b_4\varphi + \frac{1}{4}\sum_{j=1}^{k} \eta_j.$$

Hence putting $b_5 = b_4^{n-k}/4^k$, we get

$$\text{(5)} \qquad \eta_{D,\lambda}^n|U \geq \lambda^{n-k}b_5\varphi^{n-k} \wedge \eta_1 \wedge \cdots \wedge \eta_k.$$

The section s_j is represented by z_j and there is a positive C^∞ function α_j such that

$$|s_j|_j^2 = z_j \bar{z}_j \alpha_j \quad \text{for} \quad j = 1, \ldots, k.$$

So it is easy to compute

$$\partial |s_j|_j^2 = \partial(z_j \bar{z}_j \alpha_j) = \bar{z}_j \alpha_j dz_j + z_j \bar{z}_j \partial \alpha_j$$
$$\bar{\partial} |s_j|_j^2 = \bar{\partial}(z_j \bar{z}_j \alpha_j) = z_j \alpha_j d\bar{z}_j + z_j \bar{z}_j \bar{\partial} \alpha_j.$$

We have $du \wedge d^c u = (\sqrt{-1}/2\pi) \partial u \wedge \bar{\partial} u$, and so by direct computation

$$(6) \qquad du_j \wedge d^c u_j = \lambda^2 |s_j|_j^{4\lambda - 2} \frac{\sqrt{-1}}{2\pi} \alpha_j (dz_j \wedge d\bar{z}_j + \zeta_j)$$

where ζ_j is a C^∞ $(1,1)$ form which vanishes on $U \cap D_j$. Combining (5) and (6), and using the fact that α_j is bounded away from zero, we get

$$\eta_{D,\lambda}^n | U \geq \lambda^{n-k} b_6 \varphi^{n-k} \wedge \prod_{j=1}^{k} \left(\lambda^2 |s_j|_j^{2\lambda - 2} \frac{\sqrt{-1}}{2\pi} (dz_j \wedge d\bar{z}_j + \zeta_j) \right)$$

$$= \lambda^{n+k} b_6 \prod_{j=1}^{k} |s_j|_j^{2\lambda - 2} \varphi^{n-k} \wedge \prod_{j=1}^{k} \frac{\sqrt{-1}}{2\pi} (dz_j \wedge d\bar{z}_j + \zeta_j)$$

$$\geq b_U \lambda^{n+k} \prod_{j=1}^{k} |s_j|_j^{2\lambda - 2} (1 + \beta) \Phi$$

where β is a function vanishing at x. Shrinking U further we can omit mentioning this function β.

So for each point of X we have found a neighborhood where $\eta_{D,\lambda}^n$ restricted to this neighborhood satisfies the inequality desired in Lemma 7.4. The full lemma now follows by the compactness of X, thus concluding the proof.

Remark: The computation as above is similar to the curvature computation in Carlson-Griffiths [C-G], but somewhat simpler, for a couple of reasons. First it is a variation with a simpler formal structure, and second it has fewer terms since the log log terms from Carlson-Griffiths are not present here. The simpler power $|s_j|_j^{2\lambda}$ is easier to differentiate. These are some of the advantages of the singular form of Ahlfors, Stoll and Wong. Originally, Wong proved only the inequality

$$\lambda^{2n}\Omega(D)_\lambda \ll \eta_{D,\lambda}^n.$$

I improved Wong's argument in order to get the key exponent λ^{k+n} by being more careful at the appropriate technical step in the proof when taking the n-th power of $\eta_{D,\lambda}$. The structure of the proof and the number which comes out are so natural (i.e. the $1+k/n$ in the second main theorem) that there is a good possibility that the exponent $1+k/n$ is best possible. Thus conjecturally the error term is determined by local considerations on the divisor, and the complexities of its singularities.

II, §8. VARIATIONS AND APPLICATIONS

In this section I give two further applications of the crucial Proposition 7.2 and of the calculus lemma.

As in dimension 1, we have the lemma on the logarithmic derivative. A higher dimensional version was given by Griffiths [**Gr**], following Nevanlinna's differential geometric method [**Ne**] p. 259, without paying attention to constants. I gave a more precise version in [**La 8**], but using the Ahlfors-Wong method, I can now give the conjectured estimate.

Theorem 8.1. *Let $f: \mathbb{C}^n \to X$ be holomorphic non-degenerate. Let Ψ be a meromorphic n-form with no zeros, and such that its polar divisor D has simple normal crossings. Assume, for simplicity, that $f(0) \notin D$ and $0 \notin \mathrm{Ram}_f$. Let*

$$f^*\Psi = L_f(z)dz_1 \wedge \cdots \wedge dz_n.$$

Define

$$\nu_f(r) = \int_{S(r)} \log^+ |L_f| \sigma.$$

Let K be a canonical divisor and assume $-K$ is ample. Then for some constants B, B' we have

$$\nu_f(r) \le \frac{n}{2} S(BT_f^{1+k/n}, \psi, r) + B'$$

for all $r \ge r_1$ outside a set of measure $\le 2b_0(\psi)$.

Proof: By assumption, $-K$ can be taken as D, which is ample. Thus D plays the role of $(0) + (\infty)$ in the one-dimensional case. The proof is then entirely similar, following exactly the same steps as in dimension one, using the crucial estimate of Proposition 7.2. Otherwise there is no change. We carry out the details for the convenience of the reader. We let

$$T_f = T_{f,D} \quad \text{and} \quad \Lambda = 1/qT_f \quad \text{for} \quad r \ge r_1,$$

where q is the number of irreducible components of D, so that we can apply Proposition 7.2 later. We let

$$\Omega(D) = \frac{\Omega}{\Pi |s_j|_j^2} \quad \text{so that} \quad \Omega(D) = h_0 \Psi \wedge \bar{\Psi},$$

where h_0 is C^∞ on X and > 0, so bounded away from zero and infinity. The holomorphic sections s_j are always selected as in the conditions preceding Proposition 7.2, so in particular $|s_j|_j \le 1/e$. Then using the notation of those conditions, we get

$$\gamma_\Lambda = \gamma_{f,\Lambda,D} = |L_f|^2 h_D^\Lambda(h_0 \circ f) \quad \text{where} \quad h_D = \prod |s_j \circ f|_j^2.$$

We let

$$u = |L_f|^{2/n} \quad \text{and} \quad v = h_D^{\Lambda/n}.$$

Then uv and $\gamma_\Lambda^{1/n}$ differ by a positive function bounded away from zero and infinity. We then obtain the same sequence of inequalities as when $N = 1$, namely:

$$\nu_f(r) = \frac{n}{2} \int_{S(r)} (\log^+ u)\sigma$$

$$= \frac{n}{2} \int_{S(r)} (\log^+ u + \log v)\sigma - \frac{n}{2} \int_{S(r)} (\log v)\sigma$$

$$= \frac{n}{2} \int_{S(r)} (\log e^{\log^+ u + \log v})\sigma - \frac{n}{2}\Lambda(r) \int_{S(r)} (\log h_D)\sigma$$

$$\leq \frac{n}{2} \log \int_{S(r)} \gamma_\Lambda^{1/n}\sigma + \frac{n}{2}\Lambda(r)m_{f,D}(r) + O(1)$$

As in Chapter I, Theorem 6.1 we use

$$e^{\log^+ u + \log v} \leq uv + 1 \leq \gamma_\Lambda + 1.$$

Since $m_{f,D} \leq T_{f,D}$ our choice of $\Lambda(r)$ shows that the term involving $\Lambda(r)m_{f,D}(r)$ is bounded. We can then use Proposition 7.2 to conclude the proof.

I thank Alexander Eremenko for drawing my attention to the paper [G-G], where Goldberg and Grinshtein obtain a very good error term for the logarithmic derivative.

Next we come to a variation in the non-equidimensional case. In these notes, I do not want to go fully into the theory of holomorphic curves

$$f\colon \mathbf{C} \to X.$$

However, it may be worth while to give one version of such a result, stemming from Griffiths-King [G-K] and Vojta [Vo I] Theorem 5.7.2. I state the result eliminating unnecessary hypotheses.

Let Y be a complex manifold, and let $f\colon \mathbf{C} \to Y$ be a non-constant holomorphic map. If η is a $(1,1)$ form on Y then we define the **height**

$$T_{f,\eta}(r) = \int_0^r \frac{dt}{t} \int_{\mathbf{D}(t)} f^*\eta.$$

We write as usual

$$f^*\eta = \gamma_f \Phi, \qquad \text{where} \quad \Phi = \frac{\sqrt{-1}}{2\pi} dz \wedge d\bar{z}.$$

Then

$$\gamma_f = |\Delta|^2 h$$

where h is C^∞. If η is positive, then $h > 0$. Indeed, write η in local coordinates $z = (z_1, \ldots, z_n)$, and let $f = (f_1, \ldots, f_n)$. Suppose w is a complex coordinate in a neighborhood of a point in the disc so that $z = f(w)$, $f(0) = 0$, and

$$f(w) = w^e(g_1, \ldots, g_n),$$

where $g_1, \ldots, g_n$ do not all vanish at the origin. Then

$$f'(w) = w^{e-1} u(w),$$

where $u(0) \neq (0, \ldots, 0)$. Hence in these coordinates,

$$\gamma_f(w) = |w^{e-1}|^2 h(w),$$

where h is C^∞ and positive. Thus we have written γ_f in the desired form locally. The zeros of γ_f define a divisor on $\mathbf{D}(R)$, and there exists a holomorphic function Δ on $\mathbf{D}(R)$ having this divisor. We then get the desired global expression.

The function Δ is holomorphic, and defines the **ramification divisor** Ram_f, which in this case is a discrete set of zeros, with multiplicities. So we have the analogous concepts that we have met previously.

Theorem 8.2. *Let Y be a complex manifold (not necessarily compact). Let η be a closed, positive $(1,1)$ form on Y and let $f: \mathbf{D}(R) \to Y$ be a holomorphic map. Suppose there is a constant $B > 0$ such that*

$$Bf^*\eta \le \operatorname{Ric} f^*\eta.$$

Assume $0 \notin \operatorname{Ram}_f$. Then

$$BT_{f,\eta}(r) + N_{f,\mathrm{Ram}}(r) \le \frac{1}{2}S(T_{f,\eta}, \psi, r) - \frac{1}{2}\log \gamma_f(0)$$

for $r \ge r_1$ outside a set of measure $\le 2b_0(\psi)$.

Proof: The standard arguments work just as in the one-dimensional case. First observe that by definition,

$$F_{\gamma_f} = T_{f,\eta}.$$

Then by Stokes' theorem,

$$T_{\operatorname{Ric} f^*\eta}(r) + N_{f,\mathrm{Ram}}(r) = \frac{1}{2}\int\limits_0^{2\pi} \log\gamma_f(re^{i\theta})\frac{d\theta}{2\pi} - \frac{1}{2}\log \gamma_f(0).$$

By assumption, $Bf^*\eta \le \operatorname{Ric} f^*\eta$, so

$$BT_{f,\eta} \le T_{\operatorname{Ric} f^*\eta}.$$

Applying the calculus lemma to the right hand side proves the theorem.

Remarks: Several features deserve emphasis in the above theorem.

First, there is no need to assume any special property about the divisor at infinity in some compactification, e.g. normal crossings. This hypothesis is completely irrelevant here.

It is important to take Y not compact in some applications. Vojta applies the theorem to the non-compactified moduli space, for instance.

The existence of the positive $(1,1)$-form is the strongest possible assumption in the direction of hyperbolicity. For a discussion of the relation to hyperbolicity, see [**La 7**], end of Chapter III, §4 and Chapter IV, Theorem 3.3 among others.

Since Y is hyperbolic in the theorem, every holomorphic map of $\mathbf{C}$ into Y is constant. Thus in some sense Vojta's formulation is empty. But by stating the theorem for a map of a disc into Y, one gets a non-empty statement, in line with previous remarks concerning the second main theorem, which applied to maps of discs, not just $\mathbf{C}$ or $\mathbf{C}^n$, into X as in the Landau-Schottky theorem. For the statement to make sense, it is then important that the error term be given with fairly explicit constants depending on various parameters as we have done here. Then among other things, the theorem gives a bound on the radius of a disc which can be mapped into X in a non-constant way.

PART TWO

NEVANLINNA THEORY
OF COVERINGS

by William Cherry

INTRODUCTION

In this second part, the results of Part One are generalized to the case of covering spaces. Chapter III is concerned with maps from branched covers over **C** into the complex projective line, and Chapter IV works with maps from analytic coverings of $\mathbf{C}^n$ into compact n-complex dimensional manifolds. Part Two is constructed to look as much like Part One as possible, and again the methods used are those of Ahlfors, Wong, Stoll, and Griffiths and King. Stoll, in particular, treats the case of maps from covers into projective linear spaces [**St**]. The goal in these notes is to keep careful track of constants in order to see exactly where the degree enters into the error term in the second main theorem. As in the first part, the equidimensional case is treated from the differential geometric, rather than the projective linear, point of view.

The motivation for these chapters comes from Vojta's dictionary relating theorems in Nevanlinna theory to number theory [**Vo 1**]. Schematically, this analogy is as follows:

Algebraic Case   **Analytic Case**

$Y \xrightarrow{f} X$ $Y \xrightarrow{f} X$

$\downarrow$ $\downarrow$

Spec **Z**   **C**

The terms involving the degree of the covering maps which appear in the error term for the second main theorem of Nevanlinna theory are analagous to the degree of an algebraic field extension in number theory. The goal is to find an expression for the error term which is uniform for all coverings. In the simpler second main theorem without a divisor, the error term will be uniform in terms of both the covering map, and the map into the compact manifold, except for an additive constant depending on the values of the maps at the points lying above zero. Thus, the error term will in fact be uniform, if all maps are taken to satisfy a normalization condition on their values and derivatives above zero. This is in line with Vojta's conjecture in the number theoretic case, bounding the height in terms of the logarithmic disriminant, uniformly for all algebraic points, not just for points of bounded degree. In fact, the error term which appears here shows precisely that the degree comes in just as a factor multiplied by a universal expression, independent of the degree. However, this is not quite what is obtained in the more general second main theorem with a divisor. Here the second main theorem can be expressed either with an error term which can be written as the degree multiplied by an expression independent of the degree, but then added to terms which are essentially like the degree multiplied by log log of the degree, or it can be expressed with the desired form for the error term, but with the inequality holding outside an exceptional set, the size of which depends on the degree of the covering.

As in Part One, in the case of $\mathbf{C}$ or $\mathbf{C}^n$, the error term here is worked out with a general type function, ψ, so that it parallels the most refined general conjectures of approximation in the number theoretic case.

Finally, I would like to thank Serge Lang for introducing me to these questions, and for suggesting that I look at the error term for coverings to investigate its uniform behavior.

William Cherry

CHAPTER III

NEVANLINNA THEORY FOR MEROMORPHIC FUNCTIONS ON COVERINGS OF C

In this chapter, the Nevanlinna theorems for meromorphic functions on branched coverings of the complex plane are developed.

III, §1. NOTATION AND PRELIMINARIES

Let $p: Y \to \mathbf{C}$ be a covering of $\mathbf{C}$. That is, let Y be a connected Riemann surface and let p be a proper surjective holomorphic map. Let:

$[Y: \mathbf{C}] =$ the degree of the covering;

$Y(r) = \{y \in Y : |p(y)| < r\};$

$Y[r] = \{y \in Y : |p(y)| \le r\};$

$Y{<}r{>} = \{y \in Y : |p(y)| = r\};$

$\sigma_Y = d^c \log |p|^2 = p^*(\dfrac{d\theta}{2\pi})$ where $d\theta$ is the usual form on $\mathbf{C}$.

In a local coordinate w,

$$\sigma_Y(w) = \frac{\sqrt{-1}}{4\pi}\left(\frac{\overline{p'(w)}}{\overline{p(w)}}d\bar{w} - \frac{p'(w)}{p(w)}dw\right) \quad \text{outside of } Y{<}0{>} \; .$$

Proposition 1.1. *Let $f: Y \to \mathbf{P}^1$ be holomorphic. Then for $r \in \mathbf{R}_{>0}$,*

$$\int_{Y<r>} d^c \log |f|^2 = \sum_{y \in Y(r)} (\mathrm{ord}_y f)$$

Proof: The statement follows by surrounding the singularities with small circles and then applying Stoke's Theorem.
QED

Corollary 1.2. *For $r \in \mathbf{R}_{>0}$,*

$$\int_{Y<r>} \sigma_Y = [Y:\mathbf{C}].$$

Proof: Applying Proposition 1.1 to p gives

$$\int_{Y<r>} \sigma_Y = \int_{Y<r>} d^c \log |p|^2 = \sum_{y \in Y<0>} (\mathrm{ord}_y p) = [Y:\mathbf{C}].$$

QED

Corollary 1.2 is the reason that the error term in the Second Main Theorem is multipled by the degree of the covering map. The concavity of the logarithm will be used to move a log out of an integral against σ_Y and this will introduce a factor involving the degree.

In Chapter I, the Poisson-Jensen (I.1.2) and the Green-Jensen (I.2.3) integral formulas are stated and proved on $\mathbf{C}$. The conditions under which the formulas hold are stated in terms of global polar coordinates. No such global coordinates exist on Y, so first a more general version of the Green-Jensen formula will be proved and the analog of the Poisson-Jensen formula will be derived as a consequence.

114

Theorem 1.3 (Green-Jensen Formula). *Let α be a C^2 function from $Y \to \mathbf{C}$ except at a discrete set of singularities Z such that $Z \cap Y<0> = \emptyset$. Assume, in addition, that the following three conditions are satisfied:*

   i) $\alpha \sigma_Y$ is absolutely integrable on $Y<r>$ for all $r > 0$.

   ii) $d\alpha \wedge \sigma_Y$ is absolutely integrable on $Y[r]$ for all r.

   iii) $\displaystyle \lim_{\varepsilon \to 0} \int_{S(Z,\varepsilon)(r)} \alpha \sigma_Y = 0$ *for all r,*

where for sufficiently small ε, $S(Z,\varepsilon)(r)$ denotes the disjoint union of "circles" of "radius" ε around the singularities $Z \cap Y[r]$. Then

$$
\text{(A)} \qquad \int_0^r \frac{dt}{t} \int_{Y<t>} d^c\alpha = \frac{1}{2} \int_{Y<r>} \alpha \sigma_Y - \frac{1}{2} \sum_{y \in Y<0>} (\mathrm{ord}_y p)\alpha(y),
$$

and

$$
\text{(B)} \qquad \int_0^r \frac{dt}{t} \int_{Y(t)} dd^c\alpha + \int_0^r \frac{dt}{t} \lim_{\varepsilon \to 0} \int_{S(Z,\varepsilon)(t)} d^c\alpha
$$

$$
= \frac{1}{2} \int_{Y<r>} \alpha \sigma_Y - \frac{1}{2} \sum_{y \in Y<0>} (\mathrm{ord}_y p)\alpha(y).
$$

Proof: Note that (B) follows from (A) because

$$
\int_{Y(t)} dd^c\alpha + \lim_{\varepsilon \to 0} \int_{S(Z,\varepsilon)(t)} d^c\alpha = \int_{Y<t>} d^c\alpha
$$

by Stoke's Theorem and then integrating against dt/t.

1) If α and β are C^2 functions, then for degree reasons

$$
d\alpha \wedge d^c\beta = d\beta \wedge d^c\alpha.
$$

2) Because Z is disjoint from $Y<0>$,

$$\lim_{\varepsilon \to 0} \int_{Y<\varepsilon>} \alpha \sigma_Y = \sum_{y \in Y<0>} (\mathrm{ord}_y p) \alpha(y).$$

Part (A) will follow by evaluating the integral

$$\frac{1}{2} \int_{Y[r]} d(\alpha \sigma_Y)$$

in two different ways. Evaluating the integral using Stoke's Theorem gives the right hand side, and using Fubini's Theorem gives the left hand side.

Applying Stoke's Theorem, one has

$$\frac{1}{2} \int_{Y[r]} d(\alpha \sigma_Y) = \frac{1}{2} \lim_{\varepsilon \to 0} \int_{Y[r] - \left(Y(\varepsilon) \cup D(Z,\varepsilon)(r) \right)} d(\alpha \sigma_Y)$$

$$[\text{Stoke's Theorem}] \quad = \frac{1}{2} \lim_{\varepsilon \to 0} \left[\int_{Y<r>} \alpha \sigma_Y - \int_{Y<\varepsilon>} \alpha \sigma_Y - \int_{S(Z,\varepsilon)(r)} \alpha \sigma_Y \right]$$

$$[\text{by 2 and iii}] \quad = \frac{1}{2} \int_{Y<r>} \alpha \sigma_Y - \frac{1}{2} \sum_{y \in Y<0>} (\mathrm{ord}_y p) \alpha(y),$$

where $D(Z,\varepsilon)(r)$ is the disjoint union of open "discs" of "radius" ε around the singularities in $Z \cap Y<r>$. On the other hand, applying

Fubini's Theorem gives:

$$\frac{1}{2} \int_{Y[r]} d(\alpha \sigma_Y) = \frac{1}{2} \int_{Y[r]} d\alpha \wedge \sigma_Y$$

[by the definition of σ_Y] $\qquad = \dfrac{1}{2} \displaystyle\int_{Y[r]} d\alpha \wedge d^c \log |p|^2$

[by 1] $\qquad = \dfrac{1}{2} \displaystyle\int_{Y[r]} d(\log |p|^2) \wedge d^c \alpha$

$$= \int_{Y[r]} d(\log |p|) \wedge d^c \alpha$$

[from Fubini's Theorem and ii] $\qquad = \displaystyle\int_0^r \frac{dt}{t} \int_{Y<t>} d^c \alpha.$

QED

Proposition 1.4. *Let $\overline{\mathbf{D}}$ denote the closed unit disk in $\mathbf{C}$. Let f be a non-constant meromorphic function on $\overline{\mathbf{D}}$ such that 0 is the only zero or pole of f in $\overline{\mathbf{D}}$. Then,*

$$\int_{\overline{\mathbf{D}}} \left| \frac{f'}{f} \right| |dz \wedge d\bar{z}|$$

is finite.

Proof: By assumption $f(z) = z^m h(z)$ where h is never zero or infinity. Hence h'/h is continuous on $\overline{\mathbf{D}}$ and therefore bounded, say by M.

117

So

$$\int_{\overline{D}} \left| \frac{f'}{f} \right| |dz \wedge d\bar{z}| = \int_{\overline{D}} \left| \frac{m}{z} + \frac{h'}{h} \right| |dz \wedge d\bar{z}|$$

$$\leq \int_{\overline{D}} \frac{|m|}{|z|} |dz \wedge d\bar{z}| + M \int_{\overline{D}} |dz \wedge d\bar{z}|$$

$$= |m| \int_0^1 \int_0^{2\pi} \frac{1}{r} r \, dr d\theta + M \int_0^1 \int_0^{2\pi} r \, dr d\theta$$

$$= 2\pi r |m| + \pi M < \infty.$$

QED

Proposition 1.5. *Let $f: Y \to \mathbf{P}^1$ be a non-constant holomorphic map such that for $y \in Y<0>$, $f(y) \neq 0, \infty$. Let $\alpha = \log|f|$. Then, α satisfies the conditons in Theorem 1.3.*

Proof:

i) Note that $\alpha\sigma_Y$ is absolutely integrable on $Y<r>$ for the same reason $\log x$ is absolutely integrable on $[-1, 1]$.

ii) Next $d\alpha \wedge \sigma_Y$ is absolutely integrable on $Y[r]$. Indeed, since $f(y) \neq 0, \infty$ for $y \in Y<0>$, there exists an $\varepsilon > 0$ such that $\overline{Y(\varepsilon)} \cap \overline{D(Z, \varepsilon)(r)} = \emptyset$. On $Y[r] - (D(Z, \varepsilon)(r) \cup Y(\varepsilon))$, $d\alpha \wedge \sigma_Y$ is bounded, f'/f is bounded on $Y(\varepsilon)$ and p'/p is bounded on $D(Z, \varepsilon)(r)$. In a local coordinate w, $d\alpha \wedge \sigma_Y$ is given by:

$$\frac{\sqrt{-1}}{2\pi} \left(\frac{f'(w)}{f(w)} \frac{\overline{p'(w)}}{\overline{p(w)}} + \frac{\overline{f'(w)}}{\overline{f(w)}} \frac{p'(w)}{p(w)} \right) dw \wedge d\bar{w}.$$

Hence

$$|d\alpha \wedge \sigma_Y| \leq \frac{1}{\pi} \left| \frac{f'(w)}{f(w)} \right| \left| \frac{p'(w)}{p(w)} \right| |dw \wedge d\bar{w}|.$$

Therefore, the absolute integrability of $d\alpha \wedge \sigma_Y$ follows from Proposition 1.4.

iii) One has

$$\lim_{\varepsilon \to 0} \int_{S(Z,\varepsilon)(r)} \alpha \sigma_Y = 0.$$

Let y_0 be a zero of f. Let w be a complex coordinate in a neighborhood of y_0 such that $w = 0$ corresponds to y_0. Then $f(w) = w^m h(w)$ in a neighborhood of y_0, where h is holomorphic and non-vanishing. Furthermore

$$\int_{|w|=\varepsilon} \log |f|^2 \sigma_Y = \int_{|w|=\varepsilon} \log |w|^{2m} \sigma_Y + \int_{|w|=\varepsilon} \log |h|^2 \sigma_Y$$

$$= \log \varepsilon^{2m} \int_{|w|=\varepsilon} \sigma_Y + \int_{|w|=\varepsilon} \log |h|^2 \sigma_Y.$$

By Stoke's Theorem,

$$\int_{|w|=\varepsilon} \sigma_Y = \int_{|w|\leq\varepsilon} d\sigma_Y = 0$$

since $y_0 \notin Y<0>$ and σ_Y is C^∞ away from $Y<0>$. But, $\log |h|^2 \sigma_Y$ is also C^∞ in a neighborhood of y_0 since h does not vanish, so

$$\lim_{\varepsilon \to 0} \int_{|w|=\varepsilon} \log |h|^2 \sigma_Y = 0.$$

Therefore

$$\lim_{\varepsilon \to 0} \int_{S(y_0,\varepsilon)} \log |f|^2 \sigma_Y = 0.$$

A similar argument shows that

$$\lim_{\varepsilon \to 0} \int_{S(y_0,\varepsilon)} \log |f|^2 \sigma_Y = 0$$

where y_0 is a pole of f.
QED

119

Proposition 1.6. *Let $f: Y \to \mathbf{P}^1$ be a non-constant holomorphic map such that for $y \in Y<0>$, $f(y) \neq 0, \infty$. Let $\alpha = \log(1 + |f|^2)$. Then, α satisfies the conditions of Theorem 1.3.*

Proof: The singularities of α are the poles of f. Let y_0 be a pole of f, and let U be a neighborhood of y_0 such that there are holomorphic functions $f_0, f_1: U \to \mathbf{C}$ such that $f = f_1/f_0$ on U. Then

$$\log(1 + |f|^2) = \log(|f_0|^2 + |f_1|^2) - \log|f_0|^2$$

But $\log(|f_0|^2 + |f_1|^2)$ is C^∞ and $\log|f_0|^2$ satisfies the conditions of Theorem 1.3 by Proposition 1.5.
QED

Theorem 1.7 (Poisson-Jensen Formula). *Let $f: Y \to \mathbf{P}^1$ be a non-constant holomorphic map such that $f(y) \neq 0, \infty$ for $y \in Y<0>$. Then*

$$\sum_{y \in Y<0>} (\mathrm{ord}_y p) \log |f(y)| = \int_{Y<r>} \log |f| \, \sigma_Y - \sum_{y \in Y(r)} (\mathrm{ord}_y f) \log \left| \frac{r}{p(y)} \right|.$$

Proof: By Proposition 1.5, Theorem 1.3 applies to $\log |f|^2$. Hence

$$\frac{1}{2} \int_{Y<r>} \log |f|^2 \sigma_Y$$

$$= \frac{1}{2} \sum_{y \in Y<0>} (\mathrm{ord}_y p) \log |f(y)|^2 + \int_0^r \frac{dt}{t} \int_{Y<t>} d^c(\log |f|^2)$$

from Theorem 1.3 (A). But

$$\int_{Y<t>} d^c(\log |f|^2) = \sum_{y \in Y(r)} (\mathrm{ord}_y f).$$

120

Hence

$$\int\limits_0^r \frac{dt}{t} \int\limits_{Y<t>} d^c \log |f|^2 = \sum_{y \in Y(r)} (\operatorname{ord}_y f) \log \left| \frac{r}{p(y)} \right|.$$

QED

III, §2. FIRST MAIN THEOREM

In this section, Nevanlinna's First Main Theorem on the independence of the height is shown to hold in the covering case as well.

Henceforth, let $f : Y \to \mathbf{P}^1$ be a non-constant holomorphic map such that $f(y) \neq 0, \infty$ and $f'(y) \neq 0$, for all $y \in Y<0>$.

Let $n_f(0, r)$ denote the number of zeros of f in $Y(r)$ counted with multiplicities.

For $a \in \mathbf{C}$, let $n_f(a, r) = n_{f-a}(0, r)$.

Let $n_f(\infty, r) = n_{1/f}(0, r)$.

For simplicity's sake, the Nevanlinna height and counting functions will only be defined for values $a \in \mathbf{P}^1$ such that $f(y) \neq a$ for $y \in Y<0>$. For $a \in \mathbf{P}^1$ such that $f(y) \neq a$, for all $y \in Y<0>$, let

$$N_f(a, r) = \int\limits_0^r \frac{dt}{t} n_f(a, t) = \sum_{y \in Y(r)} (\operatorname{ord}_y f) \log \left| \frac{r}{p(y)} \right|.$$

For $a \in \mathbf{P}^1$, define

$$m_f(a, r) = \int\limits_{Y<r>} -\log \|f, a\| \sigma_Y,$$

where $\| \, , \, \|$ is the "chordal distance" on $\mathbf{P}^1$ defined in Chapter I, §1. Finally, for $a \in \mathbf{P}^1$ such that $f(y) \neq a$ for $y \in Y<0>$, define

$$T_{f,a}(r) = m_f(a, r) + N_f(a, r) + \sum_{y \in Y<0>} (\operatorname{ord}_y p) \log \|f(y), a\|.$$

Theorem 2.1 (First Main Theorem). $T_{f,a}(r)$ *is independenet of* $a \in \mathbf{P}^1$ *provided* $f(y) \neq a$ *for* $y \in Y<0>$.

Proof: Let $a \in \mathbf{C}$ such that $f(y) \neq a$, for all $y \in Y<0>$. From the definitions of the symbols involved:

$$
\begin{aligned}
T_{f,a}(r) = \quad & N_f(a,r) + m_f(a,r) + \sum_{y \in Y<0>} (\mathrm{ord}_y p) \log \|f(y), a\| \\
= & N_{f-a}(0,r) - \int_{Y<r>} \log \|f, a\| \sigma_Y \\
& + \sum_{y \in Y<0>} (\mathrm{ord}_y p) \log \|f(y), a\| \\
= & N_{f-a}(0,r) - \int_{Y<r>} \log |f - a| \sigma_Y \\
& + \frac{1}{2} \int_{Y<r>} \log(1 + |f|^2) \sigma_Y \\
& + \frac{1}{2} \int_{Y<r>} \log(1 + |a|^2) \sigma_Y \\
& + \sum_{y \in Y<0>} (\mathrm{ord}_y p) \log |f(y) - a| \\
& - \frac{1}{2} \sum_{y \in Y<0>} (\mathrm{ord}_y p) \log(1 + |f(y)|^2) \\
& - \frac{1}{2} \sum_{y \in Y<0>} (\mathrm{ord}_y p) \log(1 + |a|^2).
\end{aligned}
$$

Now

$$
\int_{Y<r>} \log(1 + |a|^2) \sigma_Y = [Y:\mathbf{C}] \log(1 + |a|^2) = \sum_{y \in Y<0>} (\mathrm{ord}_y p) \log(1 + |a|^2),
$$

122

and from Theorem 1.7

$$\int_{Y<r>} \log|f - a|\sigma_Y =$$

$$\sum_{y \in Y<0>} (\text{ord}_y p) \log |f(y) - a| + N_{f-a}(0, r) - N_{f-a}(\infty, r).$$

Furthermore

$$N_{f-a}(\infty, r) = N_f(\infty, r) \quad \text{and} \quad \log \|f(y), \infty\| = -\frac{1}{2}\log(1 + |f(y)|^2).$$

Hence

$$T_{f,a}(r) = \int_{Y<r>} -\log \|f, \infty\| + N_f(\infty, r)$$

$$+ \sum_{y \in Y<0>} (\text{ord}_y p) \log \|f(y), \infty\|$$

$$= m_f(\infty, r) + N_f(\infty, r) + \sum_{y \in Y<0>} (\text{ord}_y p) \log \|f(y), \infty\|$$

$$= T_{f,\infty}(r).$$

QED

In light of Theorem 2.1, denote $T_{f,\infty}(r)$ by $T_f(r)$.

The **Ahlfors-Shimizu** expression for the height in the covering case is identical to the complex plane case. Let

$$\Phi_Y = p^* \cdot \left(\frac{\sqrt{-1}}{2\pi} dz \wedge d\bar{z} \right) = dd^c |p|^2$$

$$= |p'(w)|^2 \frac{\sqrt{-1}}{2\pi} dw \wedge d\bar{w} = d|p|^2 \wedge \sigma_Y$$

be the pseudo-volume form obtained by pulling back the Euclidean volume form on **C**.

Let

$$\gamma_f = \frac{|f'|^2}{(1 + |f|^2)^2 |p'|^2}.$$

123

Note the appearance of $|p'|^2$ in the denominator of γ_f will exactly cancel the same term which appears in Φ_Y when expressed in a local coordinate.

Theorem 2.2 (Ahlfors-Shimizu). *Away from singularites, one has*

$$\gamma_f \Phi_Y = dd^c \log(1 + |f|^2) = -\frac{1}{2} dd^c \log \gamma_f,$$

and hence

$$T_f(r) = \int_0^r \frac{dt}{t} \int_{Y(t)} \gamma_f \Phi_Y$$

$$= \int_0^r \frac{dt}{t} \int_{Y(t)} dd^c \log(1 + |f|^2)$$

$$= -\frac{1}{2} \int_0^r \frac{dt}{t} \int_{Y(t)} dd^c \log \gamma_f$$

Proof: The first statement follows from the fact that

$$dd^c \log |f'|^2 = dd^c \log |p'|^2 = 0$$

away from singularities, and from the fact that $dd^c \log$ transforms products into sums.

By Proposition 1.6, Theorem 1.3 (B) can be applied to $\log(1 + |f|^2)$

to conclude:

$$\int_0^r \frac{dt}{t} \int_{Y(t)} dd^c \log(1 + |f|^2) = \frac{1}{2} \int_{Y<r>} \log(1 + |f|^2)\sigma_Y$$

$$- \frac{1}{2} \sum_{y \in Y<0>} (\mathrm{ord}_y p) \log(1 + |f(y)|^2)$$

$$- \int_0^r \frac{dt}{t} \lim_{\varepsilon \to 0} \int_{S(Z,\varepsilon)(t)} d^c \log(1 + |f|^2).$$

Now

$$\frac{1}{2} \int_{Y<r>} \log(1 + |f|^2)\sigma_Y = \int_{Y<r>} -\log \|f, \infty\| \sigma_Y = m_f(\infty, r),$$

and

$$-\frac{1}{2} \sum_{y \in Y<0>} (\mathrm{ord}_y p) \log(1 + |f(y)|^2) = \sum_{y \in Y<0>} (\mathrm{ord}_y p) \log \|f(y), \infty\|.$$

Let y_0 be a pole of f, and let $f = f_1/f_0$ in a neighborhood of y_0. Because $\log(|f_0|^2 + |f_1|^2)$ is C^∞, one gets

$$\lim_{\varepsilon \to 0} \int_{S(y_0,\varepsilon)} d^c \log(1 + |f|^2) = \lim_{\varepsilon \to 0} \int_{S(y_0,\varepsilon)} d^c \log(|f_0|^2 + |f_1|^2)$$

$$- \lim_{\varepsilon \to 0} \int_{S(y_0,\varepsilon)} d^c \log |f_0|^2 = (\mathrm{ord}_{y_0} f).$$

Therefore, since the singularities of $\log(1 + |f|^2)$ are precisely the poles of f, one has

$$\lim_{\varepsilon \to 0} \int_{S(Z,\varepsilon)(t)} d^c \log(1 + |f|^2) = -n_f(\infty, t).$$

125

This implies

$$-\int_0^r \frac{dt}{t} \lim_{\varepsilon \to 0} \int_{S(Z,\varepsilon)(t)} d^c \log(1 + |f|^2) = N_f(\infty, r).$$

Therefore

$$\int_0^r \frac{dt}{t} \int_{Y(t)} dd^c \log(1 + |f|^2) = m_f(\infty, r) + N_f(\infty, r)$$

$$+ \sum_{y \in Y<0>} (\mathrm{ord}_y p) \log \| f(y), \infty \| = T_f(r).$$

QED

§3. CALCULUS LEMMAS

The calculus lemmas of Chapter I, §3 apply to the covering case without change, with one exception, and here the change is merely a change in vocabulary necessitated by the absence of polar coordinates on Y.

Given a function α on Y, define the **height transform**:

$$F_\alpha(r) = \int_0^r \frac{dt}{t} \int_{Y(t)} \alpha \Phi_Y$$

for $r > 0$. For example, $T_f = F_{\gamma_f}$, so T_f is a height transform.

Let α be a function on Y such that the following conditions are satisfied:

(a) α is continuous and > 0 except at a discrete set of points.
(b) For each r, the integral $\int_{Y<r>} \alpha \sigma_Y$ is absolutely convergent and $r \mapsto \int_{Y<r>} \alpha \sigma_Y$ is a continuous function of r.
(c) There is an $r_1 \geq 1$ such that $F_\alpha(r_1) \geq e$.

Note: F_α has positive derivative, so is strictly increasing.

Lemma 3.1. *If α satisfies* (a), (b) *and* (c) *above, then F_α is C^2 and*

$$\frac{1}{r}\frac{d}{dr}\left(r\frac{dF_\alpha}{dr}\right) = 2\int_{Y<r>}\alpha\sigma_Y.$$

Proof:

$$\text{Since } F_\alpha = \int_0^r \frac{dt}{t}\int_{Y(t)}\alpha\Phi_Y \quad \text{one has} \quad \frac{dF_\alpha}{dr} = \frac{1}{r}\int_{Y(r)}\alpha\Phi_Y.$$

Therefore

$$r\frac{dF_\alpha}{dr} = \int_{Y(r)}\alpha\Phi_Y = \int_{Y(r)} d|p|^2 \wedge \alpha\sigma_Y$$

$$[\text{Fubini's Theorem}] \quad = 2\int_0^r tdt\int_{Y<t>}\alpha\sigma_Y.$$

Hence

$$\frac{d}{dr}\left(r\frac{dF_\alpha}{dr}\right) = 2r\int_{Y<r>}\alpha\sigma_Y \quad \text{and} \quad \frac{1}{r}\frac{d}{dr}\left(r\frac{dF_\alpha}{dr}\right) = 2\int_{Y<r>}\alpha\sigma_Y.$$

QED

Lemma 3.2. *If α satisfies* (a),(b) *and* (c), *then*

$$\log\int_{Y<r>}\alpha\sigma_Y \leq S(F_\alpha, b_1(F_\alpha), \psi, r)$$

for all $r \geq r_1(F_\alpha)$ outside a set of measure $\leq 2b_0(\psi)$.

Proof: Apply Lemma 3.1 and Lemma I.3.2.
QED

III, §4. RAMIFICATION AND THE SECOND MAIN THEOREM

In this section the ramification terms are defined, and an error term in Nevanlinna's Second Main Theorem is established, which is both uniform (after a normalization of the values above zero) for all non-constant holomorphic maps $f: Y \to \mathbf{P}^1$ and for all coverings $p: Y \to \mathbf{C}$.

Let $f: Y \to \mathbf{P}^1$ be a non-constant holomorphic map such that $f(y) \neq 0, \infty$ and $f'(y) \neq 0$ for all $y \in Y<0>$. Let $y_0 \in Y$, and let $f = f_1/f_0$ in a neighborhood of y_0. The **ramification index** of f at y_0 is defined to be

$$\mathbf{n}_{f,\mathrm{Ram}}(y_0) = (\mathrm{ord}_{y_0}(f_0 f_1' - f_0' f_1)).$$

Define

$$n_{f,\mathrm{Ram}}(t) = \sum_{y \in Y(t)} n_{f,\mathrm{Ram}}(y) \quad \text{and} \quad N_{f,\mathrm{Ram}}(r) = \int_0^r \frac{dt}{t} n_{f,\mathrm{Ram}}(t).$$

Note that since p is holomorphic, $N_{p,\mathrm{Ram}}(r) = N_{p'}(0, r)$.

Theorem 4.1. *Let $p: Y \to \mathbf{C}$ be a proper surjective holomorphic map such that $p'(y) \neq 0$ for all $y \in Y<0>$. Let $f: Y \to \mathbf{P}^1$ be a non-constant holomorphic map such that $f(y) \neq 0, \infty$ and $f'(y) \neq 0$ for all $y \in Y<0>$. Then, one has*

$$-2T_f(r) + N_{f,\mathrm{Ram}}(r) - N_{p,\mathrm{Ram}}(r) + \frac{1}{2}[Y:\mathbf{C}]\log[Y:\mathbf{C}]$$

$$\leq \frac{1}{2}[Y:\mathbf{C}]S(T_f, \psi, r) - \frac{1}{2}\sum_{y \in Y<0>} \log \gamma_f(y)$$

for all $r \geq r_1(T_f)$ outside a set of measure $\leq 2b_0(\psi)$, where

$$\gamma_f = \frac{|f'|^2}{(1 + |f|^2)^2 |p'|^2}.$$

Remark: The term on the right in the above inequality involving $\log \gamma_f$ depends only on the values of f, f' and p' above zero, so the right hand side is uniform in f and in p if they are normalized above zero. Furthermore, the term with $[Y:C]\log[Y:C]$ is positive and therefore actually improves the inequality.

Proof:

1) From Theorem 2.2

$$\int_0^r \frac{dt}{t} \int_{Y(t)} dd^c \log \gamma_f = -2T_f(r).$$

2) Let y_0 be a ramification point of f, and let $f = f_1/f_0$ in a neighborhood U of y_0, where $f_0, f_1 \colon U \to C$ are holomorphic without common zeros. Then, on U

$$\frac{|f'|^2}{(1+|f|^2)^2} = \frac{|f_0 f_1' - f_0' f_1|^2}{(|f_0|^2 + |f_1|^2)^2}.$$

Therefore

$$\lim_{\varepsilon \to 0} \int_{S(y_0,\varepsilon)} d^c \log \frac{|f'|^2}{(1+|f|^2)^2}$$

$$= \lim_{\varepsilon \to 0} \int_{S(y_0,\varepsilon)} d^c \log |f_0 f_1' - f_0' f_1|^2$$

$$- 2 \lim_{\varepsilon \to 0} \int_{S(y_0,\varepsilon)} d^c \log (|f_0|^2 + |f_1|^2)$$

$$= (\mathrm{ord}_{y_0}(f_0 f_1' - f_0' f_1)).$$

Hence

$$\int_0^r \frac{dt}{t} \lim_{\varepsilon \to 0} \int_{S(Z,\varepsilon)(t)} d^c \log \frac{|f'|^2}{(1+|f|^2)^2} = N_{f,\text{Ram}}(r),$$

where Z is the set of singularities.

3) Adding 1 and 2 to the identity:

$$N_{p,\text{Ram}}(r) = \int_0^r \frac{dt}{t} \lim_{\varepsilon \to 0} \int_{S(Z,\varepsilon)(t)} d^c \log |p'|^2$$

gives

$$N_{f,\text{Ram}}(r) - N_{p,\text{Ram}}(r) - 2T_f(r)$$

$$= \int_0^r \frac{dt}{t} \int_{Y(t)} dd^c \log \gamma_f + \int_0^r \frac{dt}{t} \lim_{\varepsilon \to 0} \int_{S(Z,\varepsilon)(t)} d^c \log \gamma_f.$$

4) However, $\log \gamma_f$ satisfies the conditions of Theorem 1.3 by applying Proposition 1.5 to $\log |f'|^2$ and $\log |p'|^2$ and Proposition 1.6 to $\log(1 + |f|^2)$. Hence, by Theorem 1.3 (B), the right hand side in 3 is equal to

$$\frac{1}{2} \int_{Y<r>} \log \gamma_f \sigma_Y - \frac{1}{2} \sum_{y \in Y<0>} \log \gamma_f(y)$$

$$= \frac{1}{2}[Y:C] \int_{Y<r>} \log \gamma_f \frac{\sigma_Y}{[Y:C]} - \frac{1}{2} \sum_{y \in Y<0>} \log \gamma_f(y)$$

$$\text{[Lemma I.3.5]} \quad \leq \frac{[Y:C]}{2} \log \left(\int_{Y<r>} \gamma_f \frac{\sigma_Y}{[Y:C]} \right) - \frac{1}{2} \sum_{y \in Y<0>} \log \gamma_f(y)$$

$$= \frac{1}{2}[Y:C] \log \left(\int_{Y<r>} \gamma_f \sigma_Y \right) - \frac{1}{2}[Y:C] \log[Y:C]$$

$$- \frac{1}{2} \sum_{y \in Y<0>} \log \gamma_f(y).$$

Notice that the degree enters into the above calculation because, in order to use Lemma I.3.5, σ_Y must be divided by the degree. This is the cause of the multiplicative factor of the degree appearing in front of the error term.

5) From Theorem 2.2, $F_{\gamma_f} = T_f$, so

$$\log\left(\int_{Y<r>} \gamma_f \sigma_Y\right) \leq S(T_f, \psi, r),$$

by Lemma 3.2 for $r \geq r_1(T_f)$ and outside a set of measure $\leq 2b_0(\psi)$. QED

III, §5. A GENERAL SECOND MAIN THEOREM

In this section, some of the curvature calculations in Chapter I, §5, are modified slightly for the covering case, and then a general second main theorem is stated and proved in two forms. One desires a theorem, as in the case without a divisor, where the error term can be expressed as the degree multiplied by an expression which is independent of the degree. This is not what is obtained here. The second main theorem, in the first form stated here, has an error term which is expressed as the degree multiplied by an expression independent of the degree, but then added to a term essentially of the form

$$[Y:C]\log\log[Y:C]$$

when the type function ψ is specialized to $(\log u)^{1+\epsilon}$. In its second form, the second main theorem is stated with an error term of the desired form, but the inequality is only valid outside an exceptional set for $r \geq r_2$, where r_2 is a number which depends on the degree.

Henceforth, the covering map $p: Y \to C$ will be assumed to be such that $p'(y) \neq 0$, for all $y \in Y<0>$ (i.e. Y is unramified above zero).

Let $f: Y \to \mathbf{P}^1$ be a non-constant holomorphic map such that $f(y) \neq 0, \infty$ and $f'(y) \neq 0$ for all $y \in Y<0>$.

Recall

$$\gamma_f = \frac{|f'|^2}{(1 + |f|^2)^2 |p'|^2}.$$

Lemma 5.1. *Let $\lambda \in \mathbf{R}_{>0}$ and let $a \in \mathbf{P}^1$. Then*

$$dd^c \|f, a\|^{2\lambda} = \left[\lambda^2 \|f, a\|^{-2(1-\lambda)} - \lambda(\lambda + 1)\|f, a\|^{2\lambda} \right] \gamma_f \Phi_Y.$$

Proof: The definition of γ_f was chosen so that in local coordinates, $\gamma_f \Phi_Y$ looks just like the case of the complex plane. Hence with the new symbols, the proof of Lemma I.5.4 proves the current lemma without modification.
QED

Lemmma 5.2. *Let $\lambda \in (0, 1)$ and $a \in \mathbf{P}^1$. Then,*

$$\lambda^2 \|f, a\|^{-2(1-\lambda)} \gamma_f \Phi_Y \leq 4dd^c \log(1 + \|f, a\|^{2\lambda}) + 12\lambda \gamma_f \Phi_Y.$$

Proof: Again, the symbols have been defined so that the proof of Lemma I.5.3 works as is.
QED

Proposition 5.3. *Let $a \in \mathbf{P}^1$ such that $f(y) \neq a$ for all $y \in Y<0>$. The function $\alpha = \log(1 + \|f, a\|^{2\lambda})$ satisfies the conditions of Theorem 1.3 and*

$$\int_0^r \frac{dt}{t} \lim_{\varepsilon \to 0} \int_{S(Z, \varepsilon)(t)} d^c \alpha = 0.$$

Proof: Since α is continuous, conditions i) and iii) of Theorem 1.3 are automatically satisfied, and since α is C^∞ away from points y such that $f(y) = a$, the remaining questions can be resolved locally. Let y_0 be a point such that $f(y_0) = a$, and let w be a complex coordinate in a neighborhood of y_0 such that $w = 0$ corresponds to the point y_0. Locally, $\|f, a\|^2$ can be written

$$\|f, a\|^2 = |w|^{2m} h(w).$$

where h is positive and C^∞. Hence, locally

$$\alpha = \log(1 + u^\lambda) \quad \text{where } u = |w|^{2m} h(w).$$

Now, let $w = \rho\, e^{i\xi}$ be local polar coordinates. Then

$$d\alpha = \frac{\lambda u^{\lambda-1} du}{1 + u^\lambda}$$

$$du = m|w|^{2(m-1)} h(w)(\bar{w}\, dw + w\, d\bar{w}) + |w|^{2m} dh$$

$$\Rightarrow \quad \left| u^{\lambda-1} du \right| \le M |w|^{2m\lambda-1} \qquad \text{for some constant } M$$

$$\Rightarrow \quad |d\alpha \wedge \sigma_Y| \le M' \frac{|w|^{2m\lambda-1}}{1 + |w|^{2m\lambda}} |dw \wedge d\bar{w}|$$

$$\Rightarrow \quad |d\alpha \wedge \sigma_Y| \le M'' \frac{\rho^{2m\lambda-1}}{1 + \rho^{2m\lambda}} \rho\, d\rho d\xi.$$

Therefore, $|d\alpha \wedge \sigma_Y|$ is absolutely integrable in a neighborhood of y_0. Hence α satisfies the conditions of Theorem 1.3.

The statement that

$$\lim_{\varepsilon \to 0} \int_{S(y_0, \varepsilon)} d^c \alpha = 0$$

follows from the proof of Proposition I.2.6.
QED

133

Now the degree enters into a calculation. The fact that the following estimate depends on the degree is precisely what prevents us from obtaining the error term in its desired form. When r is large compared with the degree, this problem can be overcome as will be seen in the proof of the second version of the second main theorem.

Proposition 5.4. *Let $a \in \mathbf{P}^1$. Let Λ be a decreasing function of r with $0 < \Lambda < 1$. Let*

$$\alpha_\Lambda = \|f, a\|^{-2(1-\Lambda)} \gamma_f.$$

Then

$$F_{\alpha_\Lambda} \leq \frac{[Y:\mathbf{C}] \log 4}{\Lambda^2} + \frac{12}{\Lambda} T_f(r).$$

Proof: Let $\lambda \in (0,1)$, and let $\alpha_\lambda = \|f, a\|^{-2(1-\lambda)} \gamma_f$. Then

$$F_{\alpha_\lambda}(r) = \int\limits_0^r \frac{dt}{t} \int\limits_{Y(t)} \|f, a\|^{-2(1-\lambda)} \gamma_f \Phi_Y$$

$$\leq \frac{4}{\lambda^2} \int\limits_0^r \frac{dt}{t} \int\limits_{Y(t)} dd^c \log(1 + \|f, a\|^{2\lambda}) + \frac{12}{\lambda} \int\limits_0^r \frac{dt}{t} \int\limits_{Y(t)} \gamma_f \Phi_Y$$

[From Lemma 5.2]

$$= \frac{4}{\lambda^2} \int\limits_0^r \frac{dt}{t} \int\limits_{Y(t)} dd^c \log(1 + \|f, a\|^{2\lambda}) + \frac{12}{\lambda} T_f(r).$$

[By Theorem 2.2]

But, by Proposition 5.3 and Theorem 1.3 (B),

$$\int_0^r \frac{dt}{t} \int_{Y(t)} dd^c \log(1 + \|f, a\|^{2\lambda})$$

$$= \frac{1}{2} \int_{Y<r>} \log(1 + \|f, a\|^{2\lambda}) \sigma_Y - \frac{1}{2} \sum_{y \in Y<0>} \log(1 + \|f(y), a\|^{2\lambda})$$

$$\leq \frac{1}{2}[Y : \mathbf{C}] \log 2.$$

Hence

$$F_{\alpha_\lambda}(r) \leq \frac{[Y : \mathbf{C}] \log 4}{\lambda^2} + \frac{12}{\lambda} T_f(r).$$

Since Λ is a decreasing function of r, one has for all $y \in Y[r]$

$$\|f(y), a\|^{-2(1 - \Lambda(|p(y)|))} \leq \|f(y), a\|^{-2(1 - \Lambda(r))}.$$

Therefore, $F_{\alpha_\Lambda}(r) \leq F_{\alpha_\lambda}(r)$ where $\lambda = \Lambda(r)$.

QED

Let $q \geq 1$ and let $a_1, ..., a_q$ be a finite set of distinct points in $\mathbf{P}^1$. Assume that $f(y) \neq a_j$ for all j and all $y \in Y<0>$.

Let:

$r_1 \geq 1$ such that $T_f(r_1) \geq e$;

Λ be a decreasing function of r with $0 < \Lambda < 1$;

$$\gamma_\Lambda = \prod_{j=1}^q \|f, a_j\|^{-2(1-\Lambda)} \gamma_f;$$

$$\alpha_\Lambda = \sum_{j=1}^q \|f, a_j\|^{-2(1-\Lambda)} \gamma_f.$$

For the convenience of the reader, Lemma I.4.2 is restated here.

Lemma 5.5. *Let:*

$$b_3 = s^{-2(q-1)} \quad where \quad s = \frac{1}{3} \min_{i \neq j} \|a_i, a_j\|$$

Then $\gamma_\Lambda \leq b_3 \alpha_\Lambda$, and b_3 depends only on $a_1, ..., a_q$.

Lemma 5.6. *Let*

$$\Lambda_1(r) = \begin{cases} \dfrac{1}{qT_f(r)} & for\ r \geq r_1 \\[2mm] constant & for\ r \leq r_1. \end{cases}$$

Let r_2 be such that $qT_f(r) > [Y:C]^{1/2}$ for all $r \geq r_2$, and let

$$\Lambda_2(r) = \begin{cases} \dfrac{[Y:C]^{1/2}}{qT_f(r)} & for\ r \geq r_2 \\[2mm] constant & for\ r \leq r_2. \end{cases}$$

Let $B = q^3 \log 4 + 12q^2$. (Note: B depends only on q.) Then

$$\log \int_{Y<r>} \gamma_{\Lambda_1} \sigma_Y \leq S([Y:C]BT_f^2, b_1(T_f), \psi, r) + \log b_3$$

for $r \geq r_1$, outside a set of measure $\leq 2b_0(\psi)$ and

$$\log \int_{Y<r>} \gamma_{\Lambda_2} \sigma_Y \leq S(BT_f^2, b_1(T_f), \psi, r) + \log b_3$$

for $r \geq r_2$, outside a set of measure $\leq 2b_0(\psi)$, where b_3 is the constant of Lemma 5.5, which depends only on $a_1, ..., a_q$.

Proof: Let $\Lambda = \Lambda_1$ or Λ_2. Then

$$\log \int_{Y<r>} \gamma_\Lambda \sigma_Y \leq \log \int_{Y<r>} \alpha_\Lambda \sigma_Y + \log b_3 \qquad \text{[By Lemma 5.5]}$$

$$\leq S(F_{\alpha_\Lambda}, b_1(F_{\alpha_\Lambda}), \psi, r) + \log b_3$$

136

for $r \geq r_1(F_{\alpha_\Lambda})$ outside a set of measure $\leq 2b_0(\psi)$. But, $\alpha_\Lambda \geq \gamma_f$ so $F_{\alpha_\Lambda} \geq T_f$, which implies that $r_1(F_{\alpha_\Lambda}) \leq r_1(T_f)$, and similarly, $\alpha_\Lambda \geq \gamma_f$ implies $b_1(F_{\alpha_\Lambda}) \leq b_1(T_f)$. Hence

$$S(F_{\alpha_\Lambda}, b_1(F_{\alpha_\Lambda}), \psi, r) \leq S(F_{\alpha_\Lambda}, b_1(T_f), \psi, r) \quad \text{for} \quad r \geq r_1(T_f).$$

By Proposition 5.4,

$$F_{\alpha_\Lambda} \leq \frac{q[Y:\mathbf{C}]\log 4}{\Lambda^2} + \frac{12q}{\Lambda} T_f.$$

For $r \geq r_1$, substituting Λ_1 for Λ, one has

$$\left([Y:\mathbf{C}]q^3 \log 4 + 12q^2\right) T_f^2 \leq [Y:\mathbf{C}]BT_f^2,$$

and for $r \geq r_2$, substituting Λ_2, one has

$$\left(q^3 \log 4 + \frac{12q^2}{[Y:\mathbf{C}]^{1/2}}\right) T_f^2 \leq BT_f^2,$$

where $B = q^3 \log 4 + 12q^2$.
QED

Theorem 5.7 (Second Main Theorem). *Let $p: Y \rightarrow \mathbf{C}$ be a proper, surjective holomorphic map such that $p'(y) \neq 0$ for all $y \in Y{<}0{>}$. Let $f: Y \rightarrow \mathbf{P}^1$ be a non-constant holomorphic map such that $f(y) \neq 0, \infty$ and $f'(y) \neq 0$ for all $y \in Y{<}0{>}$. Let:*

$$\delta(Y/\mathbf{C}) = \frac{1}{2}[Y:\mathbf{C}]\log[Y:\mathbf{C}] - [Y:\mathbf{C}]^{1/2};$$

$$S_1(F, c, \psi, r) = \log \psi(F(r)) + \log \psi(crF(r)\psi(F(r)));$$

$$B = q^3 \log 4 + 12q^2;$$

$$b_3 = \text{ the constant of Lemma 5.5};$$

$$\gamma_f = \frac{|f'|^2}{\left(1 + |f|^2\right)^2 |p'|^2}.$$

137

Then, for all $r \geq r_1$ outside a set of measure $\leq 2b_0(\psi)$ and for all $b_1 \geq b_1(T_f)$, one has (first version)

$$(q-2)T_f(r) - \sum_{j=1}^{q} N_f(a_j, r) + N_{f,\mathrm{Ram}}(r) - N_{p,\mathrm{Ram}}(r)$$

$$\leq \frac{1}{2}[Y:C]\{\log(BT_f^2) + S_1([Y:C]BT_f^2, b_1, \psi, r)\}$$

$$+ \frac{1}{2}[Y:C]\log b_3 - \frac{1}{2}\sum_{y \in Y<0>} \log \gamma_f(y) + 1.$$

Furthermore, for all $r \geq r_2$ outside of a set of measure $\leq 2b_0(\psi)$ and for all $b_1 \geq b_1(T_f)$, one has (second version)

$$(q-2)T_f(r) - \sum_{j=1}^{q} N_f(a_j, r) + N_{f,\mathrm{Ram}}(r) - N_{p,\mathrm{Ram}}(r)$$

$$\leq \frac{1}{2}[Y:C]S(BT_f^2, b_1, \psi, r) + \frac{1}{2}[Y:C]\log b_3$$

$$- \frac{1}{2}\sum_{y \in Y<0>} \log \gamma_f(y) - \delta(Y/C).$$

Remark: The constant B depends only on q, and the term

$$\frac{1}{2}[Y:C]\log b_3 - \frac{1}{2}\sum_{y \in Y<0>} \log \gamma_f(y)$$

depends on the degree, the points $a_1, ..., a_q$, and the values of f, f', and p' at the points in Y which lie above zero. Furthermore, the term $\delta(Y/C)$, which appears in the second version, is bounded from below by -1 and is positive when $[Y:C] \geq 4$, in which case the inequality is improved. A third version of the second main theorem could be stated in which the size of the exceptional set depends on the degree. The S_1 terms in the first version, since they all involve the function ψ, can be viewed as correcting for the fact that the size of the exceptional set does not grow with the degree, as in the second version. It is not known whether the dependence on the degree in the S_1 terms is necessary in the sharpest form of the inequality possible if the exceptional set is to remain uniform in size for all coverings.

Proof: From Theorem 2.2,

$$T_f(r) = \int_0^r \frac{dt}{t} \int_{Y(t)} dd^c \log(1 + |f|^2).$$

Since $f - a_j$ is holomorphic,

$$\int_0^r \frac{dt}{t} \int_{Y(t)} dd^c \log |f - a_j|^2 = 0.$$

Also

$$N_f(a_j, r) - N_f(\infty, r) = \int_0^r \lim_{\varepsilon \to 0} \int_{S(Z,\varepsilon)(t)} d^c \log |f - a_j|^2,$$

and by the proof of Theorem 2.2,

$$N_f(\infty, r) = -\int_0^r \frac{dt}{t} \lim_{\varepsilon \to 0} \int_{S(Z,\varepsilon)(t)} d^c \log(1 + |f|^2).$$

Adding the above and using the definition of $\| \ , \ \|$, one gets

$T_f(r) - N_f(a_j, r) =$

$$-\int_0^r \frac{dt}{t} \int_{Y(t)} dd^c \log \|f, a_j\|^2 - \int_0^r \frac{dt}{t} \lim_{\varepsilon \to 0} \int_{S(Z,\varepsilon)(t)} d^c \log \|f, a_j\|^2.$$

If λ is a constant, one has the following:

$(1 - \lambda)T_f(r) - (1 - \lambda)N_f(a_j, r)$

$$= -\int_0^r \frac{dt}{t} \int_{Y(t)} dd^c \log \|f, a_j\|^{2(1-\lambda)} - \int_0^r \frac{dt}{t} \lim_{\varepsilon \to 0} \int_{S(Z,\varepsilon)(t)} d^c \log \|f, a_j\|^{2(1-\lambda)}.$$

where $Z = \{y \in Y : f(y) = a_j\}$. Applying Theorem 4.1 and its proof to the above yields

$$q(1 - \lambda)T_f(r) - (1 - \lambda)\sum_{j=1}^{q} N_f(a_j, r) + N_{f,\mathrm{Ram}}(r) - N_{p,\mathrm{Ram}}(r) - 2T_f(r$$

$$= \int_0^r \frac{dt}{t} \int_{Y(t)} dd^c \log \gamma_\lambda + \int_0^r \frac{dt}{t} \lim_{\epsilon \to 0} \int_{S(Z,\epsilon)(t)} d^c \log \gamma_\lambda$$

$$= \frac{1}{2} \int_{Y<r>} \log \gamma_\lambda \sigma_Y - \frac{1}{2} \sum_{y \in Y<0>} \log \gamma_\lambda(y)$$

$$\leq \frac{1}{2} \int_{Y<r>} \log \gamma_\lambda \sigma_Y - \frac{1}{2} \sum_{y \in Y<0>} \log \gamma_f(y)$$

because $-\log \|\,,\,\|^{-2(1-\Lambda)} \leq 0$ since $\|\,,\,\| \leq 1$. But r remains constant in the last integral in the above inequality, so one can replace λ by one of the functions Λ_1 or Λ_2. Now the degree enters into the calculation again; using Lemma I.3.5 and putting $\Lambda = \Lambda_1$ or Λ_2, one has

$$\frac{1}{2} \int_{Y<r>} \log \gamma_\Lambda \sigma_Y - \sum_{y \in Y<0>} \log \gamma_f(y)$$

$$\leq \frac{[Y:C]}{2} \log \left(\int_{Y<r>} \gamma_\Lambda \sigma_Y \right) - \frac{[Y:C]}{2} \log[Y:C] - \frac{1}{2} \sum_{y \in Y<0>} \log \gamma_f(y).$$

Then, by Lemma 5.6

$$\frac{1}{2}[Y:C] \log \left(\int_{Y<r>} \gamma_{\Lambda_1} \sigma_Y \right) \leq \frac{1}{2}[Y:C]S([Y:C]BT_f^2, b_1, \psi, r) + \frac{1}{2}[Y:C] \log b_3$$

for $r \geq r_1$ outside a set of measure $\leq 2b_0(\psi)$, and

$$\frac{1}{2}[Y:C] \log \left(\int_{Y<r>} \gamma_{\Lambda_2} \sigma_Y \right) \leq \frac{1}{2}[Y:C]S(BT_f^2, b_1, \psi, r) + \frac{1}{2}[Y:C] \log b_3$$

for $r \geq r_2$ outside a set of measure $\leq 2b_0(\psi)$. Now, since $-1 \leq -(1-\Lambda)$ the term in front of $\sum N_f(a_j, r)$ can be replaced by -1.

Substituting for Λ_1 when $r \geq r_1$, one has

$$q\left(1 - \frac{1}{qT_f(r)}\right)T_f(r) - \sum_j N_f(a_j, r) + N_{f,\text{Ram}}(r) - N_{p,\text{Ram}}(r)$$

$$-2T_f(r) \leq \frac{1}{2}[Y:C]S([Y:C]BT_f^2, b_1, \psi, r) + \frac{1}{2}[Y:C]\log b_3$$

$$-\frac{1}{2}[Y:C]\log[Y:C] - \frac{1}{2}\sum_{y \in Y<0>}\log \gamma_f(y).$$

Cancelling the $[Y:C]\log[Y:C]$ terms leaves:

$$(q-2)T_f(r) + N_{f,\text{Ram}}(r) - N_{p,\text{Ram}}(r) - 1 - \sum_j N_f(a_j, r)$$

$$\leq \frac{1}{2}[Y:C](\log(BT_f^2) + S_1([Y:C]BT_f^2, b_1, \psi, r))$$

$$+\frac{1}{2}[Y:C]\log b_3 - \frac{1}{2}\sum_{y \in Y<0>}\log \gamma_f(y)$$

which concludes the proof of the first version.

On the other hand, substituting for Λ_2 when $r \geq r_2$, one has

$$q\left(1 - \frac{[Y:C]^{1/2}}{qT_f(r)}\right)T_f(r) - \sum_j N_f(a_j, r) + N_{f,\text{Ram}}(r) - N_{p,\text{Ram}}(r)$$

$$-2T_f(r) \leq \frac{1}{2}[Y:C]S(BT_f^2, b_1, \psi, r) + \frac{1}{2}[Y:C]\log b_3$$

$$-\frac{1}{2}[Y:C]\log[Y:C] - \frac{1}{2}\sum_{y \in Y<0>}\log \gamma_f(y).$$

which leaves:

$$(q - 2)T_f(r) + N_{f,\text{Ram}}(r) - N_{p,\text{Ram}}(r) - [Y:C]^{1/2} - \sum_j N_f(a_j, r)$$

$$\leq \frac{1}{2}[Y:C]S(BT_f^2, b_1, \psi, r) + \frac{1}{2}[Y:C]\log b_3$$

$$- \frac{1}{2}[Y:C]\log[Y:C] - \frac{1}{2}\sum_{y \in Y<0>} \log \gamma_f(y),$$

proving the second version of the theorem.

QED

CHAPTER IV

EQUIDIMENSIONAL NEVANLINNA THEORY
ON COVERINGS OF $\mathbf{C}^n$

As is the case with maps from $\mathbf{C}^n$, the equidimensional covering case is essentially the same as the one dimensional case, once the additional necessary language for working in higher dimensions has been added. One new feature which appears here is that the covering spaces are not assumed to be manifolds, but only analytic spaces. However, as integration is defined over the regular part of the cover, this introduces only one minor technical difficulty involving the absolute integrability of functions on the cover. Second, the determinant and trace formulas show up here, much like they did in the $\mathbf{C}^n$ case, only this time the formulas are a bit more complicated.

IV, §1. NOTATION AND PRELIMINARIES

Let $p: Y \to \mathbf{C}^n$ be a normal analytic covering of $\mathbf{C}^n$. Therefore (Y, p) satisfies the following conditions:

a) Y is a connected locally compact Hausdorff space;

b) p is a proper, surjective, continuous map such that for all $z \in \mathbf{C}^n$, $p^{-1}(z)$ is a finite set of points;

c) The set of singularities, denoted Y_{sing}, in Y is of complex codimension ≥ 2.

The subset, $Y_{reg} = Y - Y_{sing}$, is called the regular part of Y, and integration over Y is defined to be integration over the regular part. The assumption that Y is normal is precisely what is needed to ensure that Stoke's Theorem can be applied. For details see Lang [**La 5**]. For simplicity, Y is assumed to be nonsingular over the origin which means that there exists an open neighborhood U of $\mathbf{C}^n$ such that $p(Y_{sing}) \cap U = \emptyset$. The Jacobian of the map p is also assumed to be non-vanishing in a neighborhood above the origin.

Let:

$[Y : \mathbf{C}^n] =$ the degree of the covering;

$z = (z_1, ..., z_n)$ be the complex coordinates of $\mathbf{C}^n$;

$$\|z\|^2 = \sum_{j=1}^n z_j \bar{z}_j;$$

$Y(r) = \{y \in Y : \|p(y)\| < r\};$

$Y[r] = \{y \in Y : \|p(y)\| \le r\};$

$Y<r> = \{y \in Y : \|p(y)\| = r\};$

Consider the following differential forms:

$\omega(z) = dd^c \log \|z\|^2;$

$\varphi(z) = dd^c \|z\|^2;$

$\sigma(z) = d^c \log \|z\|^2 \wedge \omega^{n-1}(z);$

$$\Phi(z) = \prod_{j=1}^n \left(\frac{\sqrt{-1}}{2\pi} dz_j \wedge d\bar{z}_j \right).$$

The pullback of these forms to Y via p will be denoted by a subscript Y:

$$\omega_Y = p^*\omega, \qquad \varphi_Y = p^*\varphi, \qquad \sigma_Y = p^*\sigma, \qquad \Phi_Y = p^*\Phi.$$

Let $\omega_{\mathbf{P}^n}$ be the Fubini-Study form on $\mathbf{P}^n$ defined in Chapter II, §2, and recall that:

$$\omega = \pi^*\omega_{\mathbf{P}^{n-1}},$$

where $\pi\colon \mathbf{C}^n - \{0\} \to \mathbf{P}^{n-1}$ is the natural map representing a point in projective space by its homogeneous coordinates.

Proposition 1.1. *For $r \in \mathbf{R}_{>0}$, one has*

$$\int_{Y<r>} \sigma_Y = [Y\colon\mathbf{C}^n].$$

Proof: Outside of $Y<0>$, one has

$$d\sigma_Y = d(p^*\sigma) = p^*(d\sigma) = p^*(\omega^n) = p^*(\pi^*(\omega_{\mathbf{P}^{n-1}}^n)) = 0,$$

with the last equality holding for degree reasons. Then, by Stoke's Theorem, $\int_{Y<r>} \sigma_Y$ is independent of r. Because Y is assumed regular above zero, let r be small enough so that $p^{-1}(0)$ can be covered by $[Y\colon\mathbf{C}^n]$ disjoint open sets U_j such that $Y[r] \subseteq \bigcup U_j$ and such that $\bigcup U_j$ does not contain any of the ramification points of p. Then, $p\colon Y(r) \cap U_j \to \mathbf{C}^n$ is biholomorphic with a neighborhood of the origin, and hence

$$\int_{Y<r>\cap U_j} p^*\sigma = \int_{S(r)} \sigma = 1.$$

Summing over U_j gives the result.
QED

Again, Proposition 1.1 is the reason that the error term in the Second Main Theorem is multiplied by the degree of the covering map. As in Chapter III, the next step is to prove the Green-Jensen integral formula.

Theorem 1.2 (Green-Jensen Formula). *Let α be a C^2 function from $Y \to \mathbf{C}$ except on a negligible set of singularities Z such*

145

that $Z \cap Y<0> = \emptyset$. *Assume, in addition, that the following three conditions are satisfied:*

i) $\alpha\sigma_Y$ *is absolutely integrable on* $Y<r>$ *for all* $r > 0$.

ii) $d\alpha \wedge \sigma_Y$ *is absolutely integrable on* $Y[r]$ *for all* r.

iii) $\displaystyle\lim_{\varepsilon \to 0} \int_{S(Z,\varepsilon)(r)} \alpha\sigma_Y = 0$ *for all* r,

where for sufficiently small ε, $S(Z,\varepsilon)(r)$ *denotes the boundry of the tubular neighborhood of radius* ε *around the singularities* $Z \cap Y[r]$, *which is regular for all but a discrete set of values* ε. *Then*

$$(A) \quad \int_0^r \frac{dt}{t} \int_{Y<t>} d^c\alpha \wedge \omega_Y^{n-1} = \frac{1}{2} \int_{Y<r>} \alpha\sigma_Y - \frac{1}{2} \sum_{y \in Y<0>} \alpha(y),$$

and

$$(B) \quad \int_0^r \frac{dt}{t} \int_{Y(t)} dd^c\alpha \wedge \omega_Y^{n-1} + \int_0^r \frac{dt}{t} \lim_{\varepsilon \to 0} \int_{S(Z,\varepsilon)(t)} d^c\alpha \wedge \omega_Y^{n-1}$$

$$= \frac{1}{2} \int_{Y<r>} \alpha\sigma_Y - \frac{1}{2} \sum_{y \in Y<0>} \alpha(y).$$

Proof: Note that (B) follows from (A) because

$$\int_{Y(t)} dd^c\alpha \wedge \omega_Y^{n-1} + \lim_{\varepsilon \to 0} \int_{S(Z,\varepsilon)(t)} d^c\alpha \wedge \omega_Y^{n-1} = \int_{Y<t>} d^c\alpha \wedge \omega_Y^{n-1}$$

by Stoke's Theorem since $d\omega_Y = 0$ and then integrating against dt/t.

1) If α and β are C^2 functions then, as in the proof of Theorem II.3.1,

$$d\alpha \wedge d^c\beta \wedge \omega_Y^{n-1} = d\beta \wedge d^c\alpha \wedge \omega_Y^{n-1}$$

146

for degree reasons.

2) Because Z is disjoint from $Y<0>$, one has

$$\lim_{\varepsilon \to 0} \int_{Y<\varepsilon>} \alpha\sigma_Y = \sum_{y \in Y<0>} \alpha(y)$$

by the proof of Proposition 1.1.

Part (A) will follow by evaluating the integral

$$\frac{1}{2} \int_{Y[r]} d(\alpha\sigma_Y)$$

in two different ways. Evaluating the integral using Stoke's Theorem gives the right hand side, and using Fubini's Theorem gives the left hand side.

Applying Stoke's Theorem, one has

$$\frac{1}{2} \int_{Y[r]} d(\alpha\sigma_Y) = \frac{1}{2} \lim_{\varepsilon \to 0} \int_{Y[r] - \left(Y(\varepsilon) \cup V(Z,\varepsilon)(r) \right)} d(\alpha\sigma_Y)$$

[Stoke's Theorem] $\displaystyle = \frac{1}{2} \lim_{\varepsilon \to 0} \left[\int_{Y<r>} \alpha\sigma_Y - \int_{Y<\varepsilon>} \alpha\sigma_Y - \int_{S(Z,\varepsilon)(r)} \alpha\sigma_Y \right]$

[by 2 and iii] $\displaystyle = \frac{1}{2} \int_{Y<r>} \alpha\sigma_Y - \frac{1}{2} \sum_{y \in Y<0>} \alpha(y),$

where $V(Z,\varepsilon)(r)$ is the tubular neighborhood of radius ε around the singularities in $Z \cap Y<r>$. On the other hand, applying Fubini's

147

Theorem gives:

$$\frac{1}{2} \int_{Y[r]} d(\alpha \sigma_Y) = \frac{1}{2} \int_{Y[r]} d\alpha \wedge \sigma_Y$$

[by the definition of σ_Y]
$$= \frac{1}{2} \int_{Y[r]} d\alpha \wedge d^c \log \|p\|^2 \wedge \omega_Y^{n-1}$$

[by 1]
$$= \frac{1}{2} \int_{Y[r]} d(\log \|p\|^2) \wedge d^c \alpha \wedge \omega_Y^{n-1}$$

$$= \int_{Y[r]} d(\log \|p\|) \wedge d^c \alpha \wedge \omega_Y^{n-1}$$

[from Fubini's Theorem and ii]
$$= \int_0^r \frac{dt}{t} \int_{Y<t>} d^c \alpha \wedge \omega_Y^{n-1}.$$

QED

Recall, that a function α on Y is **admissible** if it is a linear combination of the following three types of functions:

I. Functions which are C^∞.

II. Functions locally of the form $\log(h|g|^2)$, where g is holomorphic, and h is C^∞ and positive.

III. Functions locally of the form $\log(1 + (|g|^2 h)^\lambda)$, where g is holomorphic, h is C^∞ and positive, and $0 < \lambda \leq 1$.

Proposition 1.3 *Admissible functions on Y which are C^∞ in a neighborhood of $Y<0>$ satisfy the conditions of Theorem 1.2.*

Proof: Functions of type I trivially satisfy all the conditions.

Since σ_Y is C^∞ away from $Y<0>$ and since the functions in question are C^∞ near $Y<0>$, all the conditions of Theorem 1.2 can be checked locally around their singularities.

Let α be a function of type II. Let y_0 be a singularity for α. Locally, α can be assumed to be of the form $\log|g|^2$, where g is holomorphic and $g(y_0) = 0$.

First, assume that y_0 is in the regular part of Y. If g can be taken as a local coordinate around y_0, say w_1, then one can assume $\alpha = \log|w_1|^2$. When α is restricted to the line $w_2 = \ldots = w_n = 0$, it satisfies the conditions of Theorem 1.2 because this is simply the one-dimensional case. Therefore, α in fact satisfies the conditions of Theorem 1.2 because it does not depend on $w_2, \ldots, w_n$. If g can not be taken as a coordinate function in a neighborhood of y_0, but its divisor has normal crossings, then there exist coordinate functions $w_1, \ldots, w_n$ in a neighborhood of y_0 such that $g(w) = w_1^{m_1} \cdots w_n^{m_n}$. In this case, α also satisfies the conditions of the theorem because the log converts products into sums and by what was said above. Finally, if the divisor of g does not have normal crossings, then by resolving the singularity y_0, see [Hi], a neighborhood U of y_0 can be covered by an analytic space $q: \tilde{U} \to U$ such that q is proper and such that locally g lifts to $\tilde{g}$, a holomorphic function whose divisor has normal crossings at each point above y_0. Then, the necessary integrals can be estimated on $\tilde{U}$ by what was said above, and since q is proper, everything is still finite down below.

The same technique is used when the point y_0 is not a regular point of Y. By first resolving the singularity of p, and then, if necessary, resolving any singularities of α, a general function of type II satisfies all the conditions of Theorem 1.2.

The statements for functions of type III are proved in an analogous fashion, because the singularities of these functions are also a divisor on Y.

QED

Proposition 1.4 *If α is an admissible function of type I or III, which is C^∞ in a neighborhood of $Y<0>$, then*

$$\lim_{\varepsilon \to 0} \int_{S(Z,\varepsilon)(t)} d^c\alpha \wedge \omega_Y^{n-1} = 0.$$

Proof: The statement is local around the singularities of α, and since ω_Y is C^∞ away from $Y<0>$, Lemma II.3.4 applies directly.
QED

Proposition 1.5 *If α is an admissible function of type II, which is C^∞ in a neighborhood of $Y<0>$, then*

$$\lim_{\varepsilon \to 0} \int_{S(Z,\varepsilon)(t)} d^c\alpha \wedge \omega_Y^{n-1} = \int_{Z(t)} \omega_Y^{n-1}.$$

Proof: Again, the question is local around the singularities of α and ω_Y is C^∞ away from $Y<0>$, so Lemma II.3.3 gives the result.
QED

Putting Theorem 1.2 together with Proposition 1.5 yields the following.

Theorem 1.6 *If α is admissible of type II with singular set Z, and C^∞ near $Y<0>$, then*

$$\int_0^r \frac{dt}{t} \int_{Y(t)} dd^c\alpha \wedge \omega_Y^{n-1} + \int_0^r \frac{dt}{t} \int_{Z(t)} \omega_Y^{n-1}$$

$$= \frac{1}{2} \int_{Y<r>} \alpha\sigma_Y - \frac{1}{2} \sum_{y \in Y<0>} \alpha(y).$$

IV, §2. FIRST MAIN THEOREM

In this section, the Nevanlinna functions are defined, and the First Main Theorem is proved in the covering case. Henceforth, let $f: Y \to X$ be a non-degenerate (i.e. not contained in any divisor on X) holomorphic map where X is a compact n-complex dimensional manifold. Let D be a divisor on X, and L_D a line bundle with a meromorhpic section s with $(s) = D$. As in Chapter II, all line bundles are holomorphic. Let ρ be a metric on L_D. Assume that $f(y) \notin D$ for all $y \in Y<0>$. (See Chapter II, §1.)

Pre height and height

If η is a $(1,1)$ form on X, then define

$$\mathbf{t}_{f,\eta}(t) = \int_{Y(t)} f^* \eta \wedge \omega_Y^{n-1} \quad \text{and} \quad T_{f,\eta}(r) = \int_0^r \mathbf{t}_{f,\eta}(t) \frac{dt}{t}$$

and similarly

$$\mathbf{t}_{f,\rho}(t) = \int_{Y(t)} f^* c_1(\rho) \wedge \omega_Y^{n-1} \quad \text{and} \quad T_{f,\rho}(r) = \int_0^r \mathbf{t}_{f,\rho}(t) \frac{dt}{t}$$

Proximity function

Let

$$m^0_{f,D,\rho}(r) = -\frac{1}{2} \int_{Y<r>} (\log |s \circ f|_\rho^2) \sigma_Y + \frac{1}{2} \sum_{y \in Y<0>} \log |s \circ f(y)|_\rho^2$$

Counting functions

Let

$$\mathbf{n}_{f,D}(t) = \int_{(f^*D)(t)} \omega_Y^{n-1} \quad \text{and} \quad N_{f,D}(r) = \int_0^r \mathbf{n}_{f,D}(t) \frac{dt}{t}$$

151

Proposition 2.1. *Let ρ and ρ' be two metrics on L, then*

$$T_{f,\rho'} = T_{f,\rho} + O(1).$$

Proof: The result follows from Theorem 1.2 as in the proof of Proposition II.4.2.
QED

Proposition 2.2. *Let η and η' be two $(1,1)$ forms on X, and assume that η is positive. Then*

$$T_{f,\eta'} = O(T_{f,\eta}).$$

Proof: The result follows from the compactness of X as in Proposition II.4.3.
QED

Theorem 2.3 (First Main Theorem). *For any metric ρ on L_D,*

$$T_{f,\rho} = N_{f,D} + m^0_{f,D,\rho}.$$

Proof: Since $\log|s \circ f|^2_\rho$ is an admissible function of type II on Y, the result is a restatement of Theorem 1.6 with new notation.
QED

Let Φ be the Euclidean volume form on $\mathbf{C}^n$ and let $\Phi_Y = p^*(\Phi)$ be the pullback to a pseudo-volume form on Y. Let Ω be a volume form on X, and let γ_f be the non-negative C^∞ function such that

$$f^*\Omega = \gamma_f \Phi_Y.$$

Note that γ_f vanishes on the ramification divisor of f, defined locally by the zeros of the Jacobian determinant of f, and is singular along the ramification divisor of p.

The volume form Ω defines a metric κ on the canonical line bundle K on X. Since,

$$f^* \text{Ric}\, \Omega = f^* c_1(\kappa) = dd^c \log \gamma_f$$

the **height associated to the volume form** Ω is defined as:

$$T_{f,\kappa} = \int_0^r \frac{dt}{t} \int_{Y(t)} dd^c \log \gamma_f \wedge \omega_Y^{n-1}$$

Henceforth, assume that the ramification divisor for f does not intersect $Y{<}0{>}$.

Applying Theorem 1.2 and Proposition 1.5 to the new notation gives the following theorem.

Theorem 2.4. *Assume that $p: Y \to \mathbf{C}^n$ is unramified above zero, and let $f: Y \to X$ be a non-degenerate holomorphic map such that the ramification divisor of f does not intersect $Y{<}0{>}$. Then*

$$T_{f,\kappa}(r) + N_{f,\text{Ram}}(r) - N_{p,\text{Ram}}(r)$$
$$= \frac{1}{2} \int_{Y{<}r{>}} (\log \gamma_f) \sigma_Y - \frac{1}{2} \sum_{y \in Y{<}0{>}} \log \gamma_f(y).$$

Again, the calculus lemmas of II, §5 apply to the covering case almost without change.

Lemma 3.1.

$$\Phi_Y = \frac{\|p\|^{2(n-1)}}{(n-1)!} \, d\|p\|^2 \wedge \sigma_Y.$$

Proof: This is simply the pullback of the statement:

$$\Phi = \frac{\|z\|^{2(n-1)}}{(n-1)!} \, d\|z\|^2 \wedge \sigma$$

on $\mathbf{C}^n$, which is verified by direct computation using the fact that

$$d\|z\|^2 \wedge d\|z\|^2 = d^c\|z\|^2 \wedge d^c\|z\|^2 = 0.$$

Indeed,

$$d\|z\|^2 \wedge \sigma = d\|z\|^2 \wedge d^c \log \|z\|^2 \wedge \left(dd^c \log \|z\|^2\right)^{n-1}$$

$$= \frac{d\|z\|^2 \wedge d^c\|z\|^2}{\|z\|^2} \wedge \left(\frac{dd^c\|z\|^2}{\|z\|^2} - \frac{d\|z\|^2 \wedge d^c\|z\|^2}{\|z\|^4}\right)^{n-1}$$

$$= \frac{1}{\|z\|^{2n}} \, d\|z\|^2 \wedge d^c\|z\|^2 \wedge \left(dd^c\|z\|^2\right)^{n-1}$$

$$= \frac{1}{\|z\|^{2n}} \, d\|z\|^2 \wedge d^c\|z\|^2 \wedge \varphi^{n-1}$$

$$= \frac{(n-1)!}{\|z\|^{2n}} \, \mathrm{tr}\left(d\|z\|^2 \wedge d^c\|z\|^2\right) \Phi$$

$$= \frac{(n-1)!}{\|z\|^{2(n-1)}} \, \Phi.$$

QED

Given a function α on Y, define the **height transform**:

$$F_\alpha(r) = \int_0^r \frac{dt}{t^{2n-1}} \int_{Y(t)} \alpha \Phi_Y$$

for $r > 0$.

Let α be a function on Y such that the following conditions are satisfied:

(a) α is continuous and > 0 except on a divisor of Y.

(b) For each r, the integral $\int_{Y<r>} \alpha \sigma_Y$ is absolutely convergent and $r \mapsto \int_{Y<r>} \alpha \sigma_Y$ is a piecewise continuous function of r.

(c) There is an $r_1 \geq 1$ such that $F_\alpha(r_1) \geq e$.

Note: F_α has positive derivative, so is strictly increasing.

Lemma 3.2 *If α satisfies (a),(b) and (c) above, then F_α is C^2 and*

$$\frac{1}{r^{2n-1}} \frac{d}{dr}(r^{2n-1} F_\alpha'(r)) = \frac{2}{(n-1)!} \int_{Y<r>} \alpha \sigma_Y.$$

Proof:

Since $F_\alpha = \int_0^r \frac{dt}{t^{2n-1}} \int_{Y(t)} \alpha \Phi_Y$ one has $\dfrac{dF_\alpha}{dr} = \dfrac{1}{r^{2n-1}} \int_{Y(r)} \alpha \Phi_Y.$

Hence

$$r^{2n-1} \frac{dF_\alpha}{dr} = \int_{Y(r)} \alpha \Phi_Y$$

$$[\text{Lemma 3.1}] \quad = \frac{1}{(n-1)!} \int_{Y(r)} \|p\|^{2(n-1)} d\|p\|^2 \wedge \alpha \sigma_Y$$

$$[\text{Fubini's Theorem}] \quad = \frac{2}{(n-1)!} \int_0^r t^{2n-1} dt \int_{Y<t>} \alpha \sigma_Y.$$

Therefore

$$\frac{d}{dr}\left(r^{2n-1}\frac{dF_\alpha}{dr}\right) = \frac{2}{(n-1)!}r^{2n-1}\int_{Y<r>}\alpha\sigma_Y.$$

QED

Lemma 3.3. *If α satisfies (a),(b) and (c) above, then*

$$\log\int_{Y<r>}\alpha\sigma_Y \leq S(F_\alpha, b_1(F_\alpha), \psi, r) + \log\frac{(n-1)!}{2}$$

for all $r \geq r_1(F_\alpha)$ outside a set of measure $\leq 2b_0(\psi)$.

Proof: Apply Lemma 3.2 and Lemma II.5.1.

QED

IV, §4. SECOND MAIN THEOREM WITHOUT A DIVISOR

Given a $(1,1)$ form η on Y, define the **trace** and **determinant** outside the ramification points of p as follows:

$$(\det(\eta))\,\Phi_Y = \frac{1}{n!}\eta^n$$
$$(n-1)!\,\mathrm{tr}(\eta)\,\Phi_Y = \eta \wedge \varphi_Y^{n-1}.$$

Lemma 4.1. *If η is a semi-positive $(1,1)$ form on Y, then*

$$(\det(\eta))^{1/n} \leq \frac{1}{n}\mathrm{tr}(\eta)$$

for the regular points in Y which are not ramification points of p.

Proof: This is a point-wise condition, and if y_0 is a regular point of Y which is not a ramification point for p, then p is biholomorphic

156

in a neighborhood of y_0. Since everything is defined via pull-back, the statement follows from the relationship on $\mathbf{C}^n$. See the remark preceeding Theorem II.6.3.
QED

Let η be a closed, positive $(1,1)$ form such that

$$\Omega = \frac{1}{n!}\eta^n.$$

Since $f^*\Omega = \gamma_f \Phi_Y$, one finds that $\gamma_f = \det(f^*\eta)$.

Proposition 4.2. *Let* $\tau_f = \operatorname{tr}(f^*\eta)$. *Then*

$$T_{f,\eta} = (n-1)! F_{\tau_f}.$$

Proof: All the symbols have been defined so that the proof is identical to that of Proposition II.6.2.
QED

Theorem 4.3. *Assume that* $p: Y \to \mathbf{C}^n$ *is unramified above zero, and let* $f: Y \to X$ *be a non-degenerate holomorphic map which is also unramified above zero. Let* $T_{f,\kappa}$ *be the height associated to the volume form* Ω *on* X. *Then*

$$T_{f,\kappa}(r) + N_{f,\mathrm{Ram}}(r) - N_{p,\mathrm{Ram}}(r) + \frac{n}{2}[Y:\mathbf{C}^n]\log[Y:\mathbf{C}^n]$$

$$\leq [Y:\mathbf{C}^n]\frac{n}{2}S(T_{f,\eta}, b_1(T_{f,\eta}/(n-1)!), \psi, r) - \frac{1}{2}\sum_{y \in Y<0>}\log\gamma_f(y)$$

for all $r \geq r_1(T_{f,\eta}/(n-1)!)$ *outside a set of measure* $\leq 2b_0(\psi)$.

Remark: The term on the right involving $\log\gamma_f$ in the above inequality depends only on the values of f, the Jacobian of f, and the Jacobian of p above zero, so the right hand side is uniform in f and p if they are normalized above zero. Furthermore, the term with $[Y:\mathbf{C}^n]\log[Y:\mathbf{C}^n]$ is positive and therefore actually improves the inequality.

Proof:

$$T_{f,\kappa}(r) + N_{f,\mathrm{Ram}}(r) - N_{p,\mathrm{Ram}}(r) + \frac{1}{2} \sum_{y \in Y<0>} \log \gamma_f(y)$$

$$= \frac{1}{2} \int_{Y<r>} (\log \gamma_f) \sigma_Y \qquad \text{[Theorem 2.4]}$$

$$= \frac{n}{2} \int_{Y<r>} \log \gamma_f^{1/n} \sigma_Y$$

$$\leq [Y:\mathbf{C}^n]\frac{n}{2} \log \int_{Y<r>} \gamma_f^{1/n} \sigma_Y - [Y:\mathbf{C}^n]\frac{n}{2} \log[Y:\mathbf{C}^n]$$

$$\text{[Lemma I.3.5]}$$

$$\leq [Y:\mathbf{C}^n]\frac{n}{2} \log \int_{Y<r>} \tau_f \sigma_Y - [Y:\mathbf{C}^n]\frac{n}{2} \log[Y:\mathbf{C}^n]$$

$$\text{[Proposition 4.1]}$$

$$\leq [Y:\mathbf{C}^n]\frac{n}{2} S(F_{\tau_f}, b_1(F_{\tau_f}), \psi, r) + \log(n-1)!$$

$$-[Y:\mathbf{C}^n]\frac{n}{2} \log[Y:\mathbf{C}^n]$$

$$\text{[Proposition 3.3]}$$

$$\leq [Y:\mathbf{C}^n]\frac{n}{2} S(T_{f,\eta}, b_1(T_{f,\eta}/(n-1)!), \psi, r) - [Y:\mathbf{C}^n]\frac{n}{2} \log[Y:\mathbf{C}^n]$$

$$\text{[Proposition 4.2]}$$

for all $r \geq r_1(F_{\tau_f}) = r_1(T_{f,\eta}/(n-1)!)$ outside a set of measure $\leq 2b_0(\psi)$.
QED

IV, §5. A GENERAL SECOND MAIN THEOREM

In this section, a more general second main theorem, involving a divisor, is stated and proved. Ideally, one whould have a theorem where the error term can be expressed as the degree multiplied by an expression which is independent of the degree. This is not quite what is obtained here. The second main theorem is stated in this section in two forms. In the first form, the error term is expressed as the degree multiplied by an expression independent of the degree added to a term essentially of the form

$$[Y:\mathbf{C}^n]\log\log[Y:\mathbf{C}^n]$$

when the type function ψ is specialized to $(\log u)^{1+\epsilon}$. In its second form, the second main theorem is stated with an error term which is of the desired form, but the inequality is only required to hold outside of an exceptional set for $r \geq r_2$, where r_2 is a number which depends on the degree.

For the rest of this section, let:

X be a compact n-complex dimensional manifold;

Ω be a volume form on X with associated metric κ on K the
   canonical line bundle;

$T_{f,\kappa} = T_{f,\mathrm{Ric}\,\Omega}$ be the associated height;

γ_f be the function such that $f^*\Omega = \gamma_f \Phi_Y$;

$D = \sum_{j=1}^q D_j$ be a divisor on X with simple normal crossings of
   complexity k;

$L_j = L_{D_j}$ the holomorphic line bundle associated to D_j with
   hermitian metric ρ_j;

η be a closed, positive $(1,1)$ form on X such that $\eta \geq c_1(\rho_j)$ for
   all j, and $\eta^n/n! \geq \Omega$;

s_j be a holomorphic section of L_j such that $(s_j) = D_j$;

Since X is compact, after possibly multiplying s_j by a constant, assume without loss of generality that

$$|s_j|_{\rho_j} \leq 1/e \leq 1.$$

For convenience, also assume that $f(y) \notin D$ for all $y \in Y<0>$, and that $Y<0>$ does not intersect the ramification divisor of f. If λ is a constant with $0 < \lambda < 1$, then define the **Ahlfors-Wong** singular volume form

$$\Omega(D)_\lambda = \left(\prod |s_j|_j^{-2(1-\lambda)} \right) \Omega,$$

and define

$$\gamma_\lambda = \prod |s_j \circ f|_j^{-2(1-\lambda)} \gamma_f.$$

Given Λ a positive decreasing function of r with $0 < \Lambda < 1$, define

$$\gamma_\Lambda = \prod |s_j \circ f|_j^{-2(1-\Lambda)} \gamma_f.$$

Note that because of the assumption $|s_j|_j \leq 1/e \leq 1$, one has $\gamma_f \leq \gamma_\Lambda$.

The next lemma uses the curvature computation of Lemma II.7.4, and is similar to Lemma II.7.3, but since a factor involving the deree enters the calculation, it is repeated here. The appearance of the degree in this lemma is the obstacle to obtaining an error term in the second main theorem that can be expressed as the degree multiplied by an expression independent of the degree.

Lemma 5.1 *Let b be the constant of Lemma II.7.4, which depends only on Ω, D and η. Then for any decreasing function Λ with $0 < \Lambda < 1$, one has*

$$F_{\gamma_\Lambda^{1/n}}(r) \leq (q+1) \frac{b^{1/n}}{n!} \frac{T_{f,\eta}(r)}{(\Lambda(r))^{k/n}} + \frac{qb^{1/n}}{2n!} \frac{[Y:\mathbf{C}^n] \log 2}{(\Lambda(r))^{1+k/n}}$$

for all r.

Proof: Let $0 < \lambda < 1$ be constant. From Lemma II.7.4, the constant b, depending only on Ω, D and η (and in particular, not λ) is such that

$$\lambda^{n+k} \Omega(D)_\lambda \leq \frac{b}{n!} \eta_{D,\lambda}^n,$$

where $\eta_{D,\lambda}$ is the $(1,1)$ form on X given by

$$\eta_{D,\lambda} = (q+1)\lambda\eta + \sum_{j=1}^q dd^c \log(1 + |s_j|_j^{2\lambda}),$$

and $\eta_{D,\lambda}$ is closed and positive outside of D. Pulling this back via f gives

$$\lambda^{n+k}\gamma_\lambda\Phi_Y \le b(\det(f^*\eta_{D,\lambda}))\Phi_Y,$$

and hence

$$\lambda^{1+k/n}\gamma_\lambda^{1/n}\Phi_Y \le b^{1/n}(\det(f^*\eta_{D,\lambda}))^{1/n}\Phi_Y$$

$$\text{[Proposition 4.1]} \qquad \le \frac{b^{1/n}}{n}(\operatorname{tr}(f^*\eta_{D,\lambda}))\Phi_Y$$

$$= \frac{b^{1/n}}{n!}f^*\eta_{D,\lambda}\wedge\varphi_Y^{n-1}.$$

Therefore

$$\frac{n!}{b^{1/n}}\lambda^{1+k/n}F_{\gamma_\lambda^{1/n}}(r)$$

$$\le \int_0^r \frac{dt}{t^{2n-1}}\int_{Y(t)}f^*\eta_{D,\lambda}\wedge\varphi_Y^{n-1}$$

$$= \int_0^r \frac{dt}{t}\int_{Y(t)}f^*\eta_{D,\lambda}\wedge\omega_Y^{n-1}$$

$$= (q+1)\lambda\int_0^r \frac{dt}{t}\int_{Y(t)}f^*\eta\wedge\omega_Y^{n-1}$$

$$+ \sum_{j=1}^q \int_0^r \frac{dt}{t}\int_{Y(t)}dd^c\log\bigl(1+|s_j\circ f|_j^{2\lambda}\bigr)\wedge\omega_Y^{n-1}$$

$$= (q+1)\lambda T_{f,\eta}(r)$$

$$+ \frac{1}{2}\sum_{j=1}^q\left[\int_{Y<r>}\log\bigl(1+|s_j\circ f|_j^{2\lambda}\bigr)\sigma_Y - \sum_{y\in Y<0>}\log\bigl(1+|s_j\circ f(y)|_j^{2\lambda}\bigr)\right]$$

[Proposition 1.4 and Theorem 1.2 (B)]

$$\le (q+1)\lambda T_{f,\eta}(r) + \frac{q\log 2}{2}[Y:\mathbf{C}^n].$$

[the expression inside the log is ≤ 2]

This proves Lemma 5.1 for a constant λ, but since Λ is a decreasing function of r, one has

$$\gamma_\Lambda(y) \leq \gamma_\lambda(y) \text{ for } \|p(y)\| \leq r,$$

where $\lambda = \Lambda(r)$.

QED

In light of Lemma 5.1, two positive decreasing functions Λ_1 and Λ_2 will be defined. Let $r_1 = r_1(F_{\gamma_f^{1/n}})$ and let

$$\Lambda_1(r) = \begin{cases} \dfrac{1}{qT_{f,\eta}(r)} & \text{for } r \geq r_1 \\ \text{constant} & \text{for } r \leq r_1. \end{cases}$$

Note that since $\eta^n/n! \geq \Omega$, one has $F_{\gamma_f^{1/n}} \leq T_{f,\eta}/n!$. Therefore $r_1(F_{\gamma_f^{1/n}}) \geq r_1(T_{f,\eta}/n!)$, and hence one has $\Lambda_1 \leq 1$. Let r_2 be such that $qT_{f,\eta}(r) > [Y:C^n]^{n/(n+k)}$ for all $r \geq r_2$. Let

$$\Lambda_2(r) = \begin{cases} \dfrac{[Y:C^n]^{n/(n+k)}}{qT_{f,\eta}(r)} & \text{for } r \geq r_2 \\ \text{constant} & \text{for } r \leq r_2 \end{cases}$$

Note that r_2 was chosen so that $\Lambda_2 < 1$.

Lemma 5.2 *Let b be the constant of Lemma 5.1 and let*

$$B = \frac{b^{1/n}}{n}((q+1)q^{k/n} + \frac{1}{2}q^{2+k/n}\log 2).$$

Then

$$F_{\gamma_{\Lambda_1}^{1/n}}(r) \leq \frac{B}{(n-1)!}[Y:C^n]T_{f,\eta}^{1+k/n}$$

for $r \geq r_1$, and

$$F_{\gamma_{\Lambda_2}^{1/n}}(r) \leq \frac{B}{(n-1)!}T_{f,\eta}^{1+k/n}$$

for $r \geq r_2$.

Proof: The first statement follows from applying Lemma 5.1 to Λ_1, and the second statement follows from applying Lemma 5.1 to Λ_2 since the numerator in Λ_2 was chosen to exactly cancel the appearance of the degree in Lemma 5.1.

QED

Lemma 5.3 *One has*

$$\log \int_{Y<r>} \gamma_{\Lambda_1}^{1/n} \sigma_Y \leq S([Y:C^n]BT_{f,\eta}^{1+k/n}, b_1, \psi, r)$$

for all $r \geq r_1$ outside a set of measure $\leq 2b_0(\psi)$, and

$$\log \int_{Y<r>} \gamma_{\Lambda_2}^{1/n} \sigma_Y \leq S(BT_{f,\eta}^{1+k/n}, b_1, \psi, r)$$

for all $r \geq r_2$, outside a set of measure $\leq 2b_0(\psi)$, where

$$B = \frac{b^{1/n}}{n}((q+1)q^{k/n} + \frac{1}{2}q^{2+k/n}\log 2)$$
$$b_1 = b_1(F_{\gamma_f^{1/n}}) \quad and \quad r_1 = r_1(F_{\gamma_f^{1/n}})$$

Proof: Let $\Lambda = \Lambda_1$ or Λ_2. Because $\gamma_\Lambda \geq \gamma_f$, one has

$$F_{\gamma_\Lambda^{1/n}} \geq F_{\gamma_f^{1/n}} \quad \text{and} \quad F'_{\gamma_\Lambda^{1/n}} \geq F'_{\gamma_f^{1/n}}$$

Hence $b_1 = b_1(F_{\gamma_f^{1/n}})$ and $r_1 = r_1(F_{\gamma_f^{1/n}})$ are such that for $r \geq r_1$,

$$F_{\gamma_\Lambda^{1/n}}(r) \geq e \quad \text{and} \quad b_1 r^{2n-1} F'_{\gamma_\Lambda^{1/n}} \geq e.$$

From Lemma 3.3, one has

$$\log \int_{Y<r>} \gamma_\Lambda^{1/n} \sigma_Y \leq S(F_{\gamma_\Lambda^{1/n}}, b_1, \psi, r) + \log \frac{(n-1)!}{2}$$

163

for $r \geq r_1$ outside an exceptional set of measure $\leq 2b_0(\psi)$. Now from Lemma 5.2, one has

$$S(F_{\gamma_{\Lambda_1}^{1/n}}, b_1, \psi, r) + \log \frac{(n-1)!}{2} \leq S([Y:\mathbf{C}^n]BT_{f,\eta}^{1+k/n}, b_1, \psi, r)$$

for $r \geq r_1$, and

$$S(F_{\gamma_{\Lambda_2}^{1/n}}, b_1, \psi, r) + \log \frac{(n-1)!}{2} \leq S(BT_{f,\eta}^{1+k/n}, b_1, \psi, r)$$

for $r \geq r_2$.
QED

Remark: The $(n-1)!$ in the definition of the trace cancels the $\log(n-1)!$ in Lemma 3.3.

Theorem 5.4. *Assume* $p: Y \rightarrow \mathbf{C}^n$ *is unramified above zero. Let* $f: Y \rightarrow X$ *be a non-degenerate holomorphic map such that* f *is unramified above zero. Let* D, η, κ, *and* ρ_j *be as in the beginning of this section, and assume that* $f(y) \notin D$ *for all* $y \in Y{<}0{>}$ *. Let:*

$$\delta(Y/\mathbf{C}^n, k) = \frac{n}{2}[Y:\mathbf{C}^n]\log[Y:\mathbf{C}^n] - [Y:\mathbf{C}^n]^{k/(n+k)};$$

$$S_1(F, c, \psi, r) = \log \psi(F(r)) + \log \psi(cr^{2n-1}F(r)\psi(F(r)));$$

$$B = \frac{b^{1/n}}{n}((q+1)q^{k/n} + \frac{1}{2}q^{2+k/n}\log 2);$$

$$b_1 = b_1(F_{\gamma_j^{1/n}}) \quad and \quad r_1 = r_1(F_{\gamma_j^{1/n}}),$$

where b *is the constant of Lemma 5.1. Then, one has (first version)*

$$T_{f,\kappa}(r) + \sum_{j=1}^{q} T_{f,\rho_j}(r) - N_{f,D}(r) + N_{f,\mathrm{Ram}}(r) - N_{p,\mathrm{Ram}}(r)$$

$$\leq \frac{n}{2}[Y:\mathbf{C}^n]\{\log(BT_{f,\eta}^{1+k/n}) + S_1([Y:\mathbf{C}^n]BT_{f,\eta}^{1+k/n}, \psi, b_1, r)\}$$

$$- \frac{1}{2} \sum_{y \in Y{<}0{>}} \log \gamma_f(y) + 1$$

for $r \geq r_1$ outside of a set of measure $\leq 2b_0(\psi)$. Furthermore, one has (second version)

$$T_{f,\kappa}(r) + \sum_{j=1}^{q} T_{f,\rho_j}(r) - N_{f,D}(r) + N_{f,\mathrm{Ram}}(r) - N_{p,\mathrm{Ram}}(r)$$

$$\leq \frac{n}{2}[Y:\mathbf{C}^n]S(BT_{f,\eta}^{1+k/n}, \psi, b_1, r) - \frac{1}{2}\sum_{y \in Y<0>} \log \gamma_f(y) - \delta(Y/\mathbf{C}^n, k)$$

for $r \geq r_2$ outside of a set of measure $\leq 2b_0(\psi)$.

Remark: Note that the constant B depends only on D and η, the term

$$\sum_{y \in Y<0>} \log \gamma_f(y)$$

depends only on the values of f, the Jacobian of f, and the Jacobian of p above zero, and the term $\delta(Y/\mathbf{C}^n, k)$, which appears in the second version of the inequality, is bounded from below by -1, and is positive when $[Y:\mathbf{C}^n] \geq 4$, in which case it actually improves the inequality. A third version of the inequality could also be stated with the size of the exceptional set growing with the degree. The relationship between these versions is best seen by noting that the term involving S_1 in the first version contains only the error terms involving the type function ψ, and, thus, in some sense this term corrects for the fact that the exceptional set is not enlarged in the first version as it is in the second (via r_2). It is not known whether the dependence on the degree can be removed from the S_1 terms in the sharpest form of the inequality if the size of the exceptional set is to remain independent of the degree.

Proof: Let λ be a constant with $0 < \lambda < 1$. Using Theorem 1.2 (B), Proposition 1.5, and the fact that $dd^c \log$ transforms products into sums, one obtains:

$$T_{f,\kappa}(r) + (1-\lambda)\sum_{j=1}^{q} T_{f,\rho_j}(r) - (1-\lambda)\sum_{j=1}^{q} N_{f,D_j}(r)$$

$$+ N_{f,\mathrm{Ram}}(r) - N_{p,\mathrm{Ram}}(r)$$

$$= \int_0^r \frac{dt}{t} \int_{Y(t)} dd^c \log \gamma_\lambda + \int_0^r \frac{dt}{t} \lim_{\varepsilon \to 0} \int_{S(Z,\varepsilon)(t)} d^c \log \gamma_\lambda$$

$$= \frac{n}{2} \int_{Y<r>} \log \gamma_\lambda^{1/n} \sigma_Y - \frac{1}{2} \sum_{y \in Y<0>} \log \gamma_\lambda(y)$$

Because of the assumption that $|s_j|_j \leq 1$, one also has

$$-\frac{1}{2} \sum_{y \in Y<0>} \log \gamma_\lambda(y) \leq -\frac{1}{2} \sum_{y \in Y<0>} \log \gamma_f(y).$$

Also, since Λ_1 and Λ_2 are constant on $Y<r>$, the function $\Lambda = \Lambda_1$ or Λ_2 can replace λ in the above equality. Furthermore, $N_{f,D_j} \geq 0$ and $-1 \leq -(1-\lambda)$, so the factor $(1-\lambda)$ in front can be deleted. When $r \geq r_1$, one has

$$-1 \leq -\Lambda_1(r) \sum_{j=1}^{q} T_{f,\rho_j}(r),$$

and when $r \geq r_2$, one has

$$-[Y:\mathbf{C}^n]^{n/(n+k)} \leq -\Lambda_2(r) \sum_{j=1}^{q} T_{f,\rho_j}(r),$$

from the definitions of Λ_1 and Λ_2, and from the fact that η was chosen so that

$$T_{f,\eta} \geq T_{f,\rho_j} \quad \text{for all } j.$$

Finally, by moving the log out of the integral, one has

$$\int_{Y<r>} \log \gamma_{\Lambda(r)}^{1/n} \sigma_Y \leq [Y:\mathbf{C}^n] \log \left(\int_{Y<r>} \gamma_{\Lambda(r)}^{1/n} \sigma_Y \right) - [Y:\mathbf{C}^n] \log[Y:\mathbf{C}^n].$$

Applying the estimate in Lemma 5.3 for Λ_2 to the term with the integral on the right and collecting terms yields the second version of the theorem. To arrive at the first version, apply the estimate in Lemma 5.3 for Λ_1 to get

$$\log \left(\int\limits_{Y<r>} \gamma_{\Lambda_1(r)}^{1/n} \sigma_Y \right) \le S([Y:\mathbf{C}^n]BT_{f,\eta}^{1+k/n}, b_1, \psi, r)$$

for $r \ge r_1$ outside a set of measure $\le 2b_0(\psi)$. But

$$S([Y:\mathbf{C}^n]BT_{f,\eta}^{1+k/n}, b_1, \psi, r)$$
$$= \log[Y:\mathbf{C}^n] + \log(BT_{f,\eta}^{1+k/n}) + S_1([Y:\mathbf{C}^n]BT_{f,\eta}^{1+k/n}, b_1, \psi, r)$$

After multiplying by $[Y:\mathbf{C}^n](n/2)$, the $\log[Y:\mathbf{C}^n]$ term in the above equality is exactly canceled by the $\log[Y:\mathbf{C}^n]$ term which came from moving the log out of the integral above. Collecting terms then yields the first version.

QED

IV, §6. A VARIATION

In this section the analog of Theorem II.8.2, a non-equidimensional statement will be proved.

In this section, let $p: Y \to \mathbf{C}$ be a branched covering of $\mathbf{C}$ which is unramified over the origin, and let Φ_Y be the pull-back to Y of the standard Euclidean volume form on $\mathbf{C}$.

Theorem 6.1. *Let M be a complex manifold which is not necessarily compact. Let η be a closed, positive $(1,1)$ form on M, and let $f: Y(R) \to M$ be a holomorphic map which is unramified above zero. Suppose there exists a constant $B > 0$ such that*

$$Bf^*\eta \le \mathrm{Ric}\, f^*\eta.$$

167

Let

$$T_{f,\eta}(r) = \int\limits_0^r \frac{dt}{t} \int\limits_{Y(t)} f^*\eta \quad and \quad f^*\eta = \gamma_f \Phi_Y.$$

Then

$$BT_{f,\eta}(r) + N_{f,\mathrm{Ram}}(r) - N_{p,\mathrm{Ram}}(r) + \frac{1}{2}[Y:C]\log[Y:C]$$

$$\leq \frac{1}{2}[Y:C]S(T_{f,\eta},\psi,r) - \frac{1}{2}\sum_{y\in Y<0>}\log\gamma_f(y)$$

for $r \geq r_1$ outside a set of measure $\leq 2b_0(\psi)$.

Proof: By definition, one has $F_{\gamma_f} = T_{f,\eta}$. By Stoke's Theorem,

$$T_{\mathrm{Ric}(f^*\eta)}(r) + N_{f,\mathrm{Ram}}(r) - N_{p,\mathrm{Ram}}(r)$$

$$= \frac{1}{2}\int\limits_{Y<r>}\log\gamma_f\sigma_Y - \frac{1}{2}\sum_{y\in Y<0>}\log\gamma_f(y).$$

But, by assumption,

$$Bf^*\eta \leq \mathrm{Ric}\ f^*\eta, \quad so \quad BT_{f,\eta} \leq T_{\mathrm{Ric}(f^*\eta)}.$$

The theorem is concluded by applying the calculus lemma to the right hand side.

QED

References

[Ad 1] W. ADAMS, *Asymptotic diophantine approximations to e*, Proc. Nat. Acad. Sci. USA **55**, (1966), 28-31.

[Ad 2] W. ADAMS, *Asymptotic diophantine approximations and Hurwitz numbers*, Amer. J. Math. **89** (1967), 1083-1108.

[A-L] W. ADAMS and S. LANG, *Some computations in diophantine approximations*, J. reine angew. Math. **220** (1965), 163-173.

[Ah] L. AHLFORS, *The theory of meromorphic curves*, Acta. Soc. Sci. Fenn. Nova Ser. A **3** (1941), 1-31.

[Br] A.D. BRYUNO, *Continued fraction expansion of algebraic numbers*, Zh. Vychisl. Mat. i Mat. Fiz. 4 nr. **2**, 211-221; translated USSR Comput. Math. and Math. Phys 4 (1964), 1-15.

[C-G] J. CARLSON and P. GRIFFITHS, *A defect relation for equidimensional holomorphic mappings between algebraic varieties*, Ann. Math. **95** (1972), 557-584.

[Ch] S.S. CHERN, *Complex analytic mappings of Riemann surfaces I*, Amer. J. Math. **LXXXII No. 2** (1960), 323-337.

[G-G] A. GOLDBERG and V. GRINSHTEIN, *The logarithmic derivative of a meromorphic function* , Mathematical Notes 19 (1976), AMS Translation, 320-323.

[Gr] P. GRIFFITHS, *Entire Holomorphic Mappings in one and Several Complex Variables*, Ann. of Math. Studies Vol 85, Princeton University Press, Princeton, NJ, 1976.

[G-K] P. GRIFFITHS and J. KING, *Nevanlinna theory and holomorphic mappings between algebraic varieties*, Acta Mathematica **130** (1973), 145-220.

[GuR] R. GUNNING and H. ROSSI, *Introduction to Holomorphic Functions of Several Variables* (Replacing *Analytic Functions of Several Complex Variables*, Prentice Hall, 1965) (coming out in 1990)

[Hi] H. HIRONAKA, *Resolution of singularities of an algebraic variety over a field of characteristic zero:* I, II, Annals of Mathematics **79** (1964), 109-326.

[Kh] A. KHINTCHINE, *Continued fractions*, Chicago University Press, 1964

[La 1] S. LANG, *Report on diophantine approximations*, Bull. Soc. Math. France **93** (1965), 177-192.

[La 2] S. LANG, *Asymptotic diophantine approximations*, Proc. NAS **55 No. 1** (1966), 31-34.

[La 3] S. LANG, *Introduction to Diophantine Approximations*, Addison Wesley, 1966.

[La 4] S. LANG, *Transcendental numbers and diophantine approximations*, Bull. AMS **77 No. 5** (1971), 635-677.

[La 5] S. LANG, *Real Analysis*, Addison Wesley, 1969; Second Edition, 1983.

[La 6] S. LANG, *Differential Manifolds*, 1972; reprint by Springer Verlag, 1985.

[La 7] S. LANG, *Introduction to Complex Hyperbolic Spaces*, Springer Verlag, 1987.

[La 8] S. LANG, *The error term in Nevanlinna theory*, Duke Math. J. (1988), 193-218.

[La 9] S. LANG, *The error term in Nevanlinna theory* II, Bull. AMS (1990) pp. 115-125.

[L-T 1] S. LANG and H. TROTTER, *Continued fractions of some algebraic numbers*, J. reine angew. Math. **255** (1972), 112-123.

[L-T 2] S. LANG and H. TROTTER, *Addendum to the above*, J. reine angew. Math. **267** (1974), 219-220.

[Ne] R. NEVANLINNA, *Analytic Functions*, Springer Verlag, 1970; (revised translation of the German edition, 1953).

[vN-T] von NEUMANN and B. TUCKERMAN, *Continued fraction expansion of* $2^{1/3}$, Math. Tables Aids Comput **9** (1955), 23-24.

[Os 1] C.F. OSGOOD, *A number theoretic-differential equations approach to generalizing Nevanlinna theory*, Indian J. of Math. **23** (1981), 1-15.

[Os 2] C.F. OSGOOD, *Sometimes effective Thue-Siegel Roth-Schmidt-Nevanlinna bounds, or better*, J. Number Theory **21** (1985), 347-389.

[RDM] R. RICHTMYER, M. DEVANEY and N. METROPOLIS, *Continued fraction expansions of algebraic numbers*, Numer. Math. **4** (1962), 68-84.

[Sc] W. SCHMIDT, *Diophantine Approximations*, Lecture Notes in Mathematics, Springer Verlag, 1980.

[Sh] S.V. SHABAT, *Distribution of Values of HolomorphicMappings*, translated from the Russian, AMS, 1985 (Russian edition 1982).

[St] W. STOLL, *The Ahlfors-Weyl theory of meromorphic maps on parabolic manifolds*, Lecture Notes in Mathematics 981, Springer Verlag, 1981.

[Vo 1] P. VOJTA, *Diophantine Approximations and Value Distribution Theory*, Lecture Notes in Mathematics 1239, Springer Verlag, 1987.

[Vo 2] P. VOJTA, *A refinement of Schmidt's subspace theorem*, Am. J. Math **111**(1989), pp. 489-518.

[Wo] P.M. WONG, *On the second main theorem of Nevanlinna theory*, Am. J. Math. **111**(1989), pp. 549-583.

Added in proof: I thank Alexander Eremenko for drawing my attention to the paper "The Logarithmic Derivative of a Meromorphic Function" by A. A. Goldberg and V. A. Grinshtein. Mathematical Notes Vol. 19, 1976, AMS translation pp. 320–323. In that paper they obtain a very good error term for the logarithmic derivative.

INDEX

Vol. 1320: H. Jürgensen, G. Lallement, H.J. Weinert (Eds.), Semigroups, Theory and Applications. Proceedings, 1986. X, 416 pages. 1988.

Vol. 1321: J. Azéma, P.A. Meyer, M. Yor (Eds.), Séminaire de Probabilités XXII. Proceedings. IV, 600 pages. 1988.

Vol. 1322: M. Métivier, S. Watanabe (Eds.), Stochastic Analysis. Proceedings, 1987. VII, 197 pages. 1988.

Vol. 1323: D.R. Anderson, H.J. Munkholm, Boundedly Controlled Topology. XII, 309 pages. 1988.

Vol. 1324: F. Cardoso, D.G. de Figueiredo, R. Iório, O. Lopes (Eds.), Partial Differential Equations. Proceedings, 1986. VIII, 433 pages. 1988.

Vol. 1325: A. Truman, I.M. Davies (Eds.), Stochastic Mechanics and Stochastic Processes. Proceedings, 1986. V, 220 pages. 1988.

Vol. 1326: P.S. Landweber (Ed.), Elliptic Curves and Modular Forms in Algebraic Topology. Proceedings, 1986. V, 224 pages. 1988.

Vol. 1327: W. Bruns, U. Vetter, Determinantal Rings. VII,236 pages. 1988.

Vol. 1328: J.L. Bueso, P. Jara, B. Torrecillas (Eds.), Ring Theory. Proceedings, 1986. IX, 331 pages. 1988.

Vol. 1329: M. Alfaro, J.S. Dehesa, F.J. Marcellan, J.L. Rubio de Francia, J. Vinuesa (Eds.): Orthogonal Polynomials and their Applications. Proceedings, 1986. XV, 334 pages. 1988.

Vol. 1330: A. Ambrosetti, F. Gori, R. Lucchetti (Eds.), Mathematical Economics. Montecatini Terme 1986. Seminar. VII, 137 pages. 1988.

Vol. 1331: R. Bamón, R. Labarca, J. Palis Jr. (Eds.), Dynamical Systems, Valparaiso 1986. Proceedings. VI, 250 pages. 1988.

Vol. 1332: E. Odell, H. Rosenthal (Eds.), Functional Analysis. Proceedings, 1986–87. V, 202 pages. 1988.

Vol. 1333: A.S. Kechris, D.A. Martin, J.R. Steel (Eds.), Cabal Seminar 81–85. Proceedings, 1981–85. V, 224 pages. 1988.

Vol. 1334: Yu.G. Borisovich, Yu. E. Gliklikh (Eds.), Global Analysis – Studies and Applications III. V, 331 pages. 1988.

Vol. 1335: F. Guillén, V. Navarro Aznar, P. Pascual-Gainza, F. Puerta, Hyperrésolutions cubiques et descente cohomologique. XII, 192 pages. 1988.

Vol. 1336: B. Helffer, Semi-Classical Analysis for the Schrödinger Operator and Applications. V, 107 pages. 1988.

Vol. 1337: E. Sernesi (Ed.), Theory of Moduli. Seminar, 1985. VIII, 232 pages. 1988.

Vol. 1338: A.B. Mingarelli, S.G. Halvorsen, Non-Oscillation Domains of Differential Equations with Two Parameters. XI, 109 pages. 1988.

Vol. 1339: T. Sunada (Ed.), Geometry and Analysis of Manifolds. Procedings, 1987. IX, 277 pages. 1988.

Vol. 1340: S. Hildebrandt, D.S. Kinderlehrer, M. Miranda (Eds.), Calculus of Variations and Partial Differential Equations. Proceedings, 1986. IX, 301 pages. 1988.

Vol. 1341: M. Dauge, Elliptic Boundary Value Problems on Corner Domains. VIII, 259 pages. 1988.

Vol. 1342: J.C. Alexander (Ed.), Dynamical Systems. Proceedings, 1986–87. VIII, 726 pages. 1988.

Vol. 1343: H. Ulrich, Fixed Point Theory of Parametrized Equivariant Maps. VII, 147 pages. 1988.

Vol. 1344: J. Král, J. Lukeš, J. Netuka, J. Veselý (Eds.), Potential Theory – Surveys and Problems. Proceedings, 1987. VIII, 271 pages. 1988.

Vol. 1345: X. Gomez-Mont, J. Seade, A. Verjovski (Eds.), Holomorphic Dynamics. Proceedings, 1986. VII, 321 pages. 1988.

Vol. 1346: O. Ya. Viro (Ed.), Topology and Geometry – Rohlin Seminar. XI, 581 pages. 1988.

Vol. 1347: C. Preston, Iterates of Piecewise Monotone Mappings on an Interval. V, 166 pages. 1988.

Vol. 1348: F. Borceux (Ed.), Categorical Algebra and its Applications. Proceedings, 1987. VIII, 375 pages. 1988.

Vol. 1349: E. Novak, Deterministic and Stochastic Error Bounds in Numerical Analysis. V, 113 pages. 1988.

Vol. 1350: U. Koschorke (Ed.), Differential Topology. Proceedings, 1987. VI, 269 pages. 1988.

Vol. 1351: I. Laine, S. Rickman, T. Sorvali, (Eds.), Complex Analysis, Joensuu 1987. Proceedings. XV, 378 pages. 1988.

Vol. 1352: L. L. Avramov, K.B. Tchakerian (Eds.), Algebra – Some Current Trends. Proceedings, 1986. IX, 240 Seiten. 1988.

Vol. 1353: R.S. Palais, Ch.-I. Terng, Critical Point Theory and Submanifold Geometry. X, 272 pages. 1988.

Vol. 1354: A. Gómez, F. Guerra, M.A. Jiménez, G. López (Eds.) Approximation and Optimization. Proceedings, 1987. VI, 280 pages 1988.

Vol. 1355: J. Bokowski, B. Sturmfels, Computational Synthetic Geometry. V, 168 pages. 1989.

Vol. 1356: H. Volkmer, Multiparameter Eigenvalue Problems and Expansion Theorems. VI, 157 pages. 1988.

Vol. 1357: S. Hildebrandt, R. Leis (Eds.), Partial Differential Equations and Calculus of Variations. VI, 423 pages. 1988.

Vol. 1358: D. Mumford, The Red Book of Varieties and Schemes. V 309 pages. 1988.

Vol. 1359: P. Eymard, J.-P. Pier (Eds.), Harmonic Analysis. Proceedings, 1987. VIII, 287 pages. 1988.

Vol. 1360: G. Anderson, C. Greengard (Eds.), Vortex Methods Proceedings, 1987. V, 141 pages. 1988.

Vol. 1361: T. tom Dieck (Ed.), Algebraic Topology and Transformation Groups. Proceedings, 1987. VI, 298 pages. 1988.

Vol. 1362: P. Diaconis, D. Elworthy, H. Föllmer, E. Nelson, G. Papanicolaou, S.R.S. Varadhan. École d'Été de Probabilités de Saint Flour XV–XVII, 1985–87. Editor: P.L. Hennequin. V, 459 pages 1988.

Vol. 1363: P.G. Casazza, T.J. Shura. Tsirelson's Space. VIII, 204 pages. 1988.

Vol. 1364: R.R. Phelps, Convex Functions, Monotone Operators and Differentiability. IX, 115 pages. 1989.

Vol. 1365: M. Giaquinta (Ed.), Topics in Calculus of Variations. Seminar, 1987. X, 196 pages. 1989.

Vol. 1366: N. Levitt, Grassmannians and Gauss Maps in PL-Topology. V, 203 pages. 1989.

Vol. 1367: M. Knebusch, Weakly Semialgebraic Spaces. XX, 376 pages. 1989.

Vol. 1368: R. Hübl, Traces of Differential Forms and Hochschild Homology. III, 111 pages. 1989.

Vol. 1369: B. Jiang, Ch.-K. Peng, Z. Hou (Eds.), Differential Geometry and Topology. Proceedings, 1986–87. VI, 366 pages. 1989.

Vol. 1370: G. Carlsson, R.L. Cohen, H.R. Miller, D.C. Ravenel (Eds.), Algebraic Topology. Proceedings, 1986. IX, 456 pages. 1989.

Vol. 1371: S. Glaz, Commutative Coherent Rings. XI, 347 pages. 1989.

Vol. 1372: J. Azéma, P.A. Meyer, M. Yor (Eds.), Séminaire de Probabilités XXIII. Proceedings. IV, 583 pages. 1989.

Vol. 1373: G. Benkart, J.M. Osborn (Eds.), Lie Algebras, Madison 1987. Proceedings. V, 145 pages. 1989.

Vol. 1374: R.C. Kirby, The Topology of 4-Manifolds. VI, 108 pages. 1989.

Vol. 1375: K. Kawakubo (Ed.), Transformation Groups. Proceedings, 1987. VIII, 394 pages, 1989.

Vol. 1376: J. Lindenstrauss, V.D. Milman (Eds.), Geometric Aspects of Functional Analysis. Seminar (GAFA) 1987–88. VII, 288 pages. 1989.

Vol. 1377: J.F. Pierce, Singularity Theory, Rod Theory, and Symmetry Breaking Loads. IV, 177 pages. 1989.

Vol. 1378: R.S. Rumely, Capacity Theory on Algebraic Curves. III, pages. 1989.

Vol. 1379: H. Heyer (Ed.), Probability Measures on Groups. Proceedings, 1988. VIII, 437 pages. 1989